AF581718

Application of Advanced Quantum Dots Films in Optoelectronics

Application of Advanced Quantum Dots Films in Optoelectronics

Editors

Philippe Guyot-Sionnest
Xin Tang

Basel • Beijing • Wuhan • Barcelona • Belgrade • Novi Sad • Cluj • Manchester

Editors
Philippe Guyot-Sionnest
The University of Chicago
Chicago, IL, USA

Xin Tang
Beijing Institute of
Technology
Beijing, China

Editorial Office
MDPI
St. Alban-Anlage 66
4052 Basel, Switzerland

This is a reprint of articles from the Special Issue published online in the open access journal *Coatings* (ISSN 2079-6412) (available at: http://www.mdpi.com).

For citation purposes, cite each article independently as indicated on the article page online and as indicated below:

Lastname, A.A.; Lastname, B.B. Article Title. *Journal Name* **Year**, *Volume Number*, Page Range.

ISBN 978-3-0365-9018-9 (Hbk)
ISBN 978-3-0365-9019-6 (PDF)
doi.org/10.3390/books978-3-0365-9019-6

Contents

Preface

Colloidal quantum dots (CQDs) and other thin-film materials have been extensively investigated over the past few decades in various fields. Their advantages, such as their solution processability, tunable energy gaps and surface-chemistry-mediated electronic properties, make CQDs a fascinating material platform to develop optoelectronic devices for both light emission and detection. Driven by the tremendous demands for energy harvesting, photodetectors and displays, synthetic strategies toward improving material properties, improving quantum efficiency through device engineering and achieving light manipulation using nanophotonic structures are all being actively investigated in CQD-based optoelectronics. The scope of this book is to provide a platform to both academic and industry researchers to share their state-of-the-art developments in the rapidly growing field of optoelectronics.

Philippe Guyot-Sionnest and Xin Tang
Editors

Editorial

Special Issue "Application of Advanced Quantum Dots Films in Optoelectronics"

Xin Tang [1,2,3]

1 School of Optics and Photonics, Beijing Institute of Technology, Beijing 100081, China; xintang@bit.edu.cn
2 Beijing Key Laboratory for Precision Optoelectronic Measurement Instrument and Technology, Beijing 100081, China
3 Yangtze Delta Region Academy of Beijing Institute of Technology, Jiaxing 314019, China

Colloidal quantum dots (CQDs) have been extensively investigated in recent decades. As an alternative to bulk semiconductors, the energy gaps of CQDs can be tuned by not only the CQDs' composition but also their size. So far, CQDs have been used in a wide range of applications including both light emission and detection. This Special Issue aims to provide a forum for researchers to share their current research findings and to promote further research into CQD-based optoelectronics, including experimental characterization and theoretical calculations. For this Special Issue, we intend to collect 10 original articles on the most recent works on CQD light-emitting diodes, infrared photodetectors, and X-ray sensors.

In the work by Zhang et al. [1], mercury telluride CQDs (HgTe CQDs) were introduced as a new option for infrared detection. Limited by their energy gaps, early studies on PbS CQDs mainly focused on solar cells or near-infrared detection. The emergence of HgTe CQDs greatly extended the spectral sensing ranges of colloidal nanomaterials to mid-infrared and THz ranges. The synthesis, device physics, photodetection mechanism, and multispectral imaging of HgTe CQDs have been discussed. Moreover, nanomaterials are subject to quantum confinement effects. Therefore, they should have a smaller energy state density, which might lead to a reduced dark current and a higher operation temperature. Novel low-dimensional materials with easy fabrication processes and excellent photoelectronic properties provide a possible solution for room temperature infrared photodetectors. Li et al. summarized the preparation methods and characterization of several low-dimensional materials (PbS, PbSe, and HgTe, new two-dimensional materials) and the room temperature infrared photodetectors based on them [2].

For traditional infrared semiconductors, focal-plane arrays (FPAs) have to be fabricated via complicated flip-bonding methods, leading to expensive infrared imagers. Emerging materials, such as inorganic–organic metal halide perovskites, organic polymers, and colloidal quantum dots, have been proposed to develop CMOS-compatible optoelectronic imagers. In the work by Bi et al. [3], the fabrication methods and key figures of merit for FPAs were discussed.

Besides imaging, spectral sensing is of great importance. Unlike HgTe CQDs, HgSe CQDs show a narrowband photoresponse via intraband transitions. Zhao et al. investigated an intraband mid-infrared photodetector based on HgSe colloidal quantum dots (CQDs) [4]. The size, absorption spectra, and carrier mobility of HgSe CQD films were all experimentally studied. Wen et al. reported a simulation study of a microspectrometer fabricated by integrating an intraband HgSe CQD detector with a distributed Bragg reflector (DBR) [5]. Intraband HgSe CQDs possess a unique narrowband absorption and optical response, which makes them an ideal material platform to achieve high-resolution detection for infrared signatures such as molecular vibration.

Integrating CQD films with functional optical structures is an effective way to provide functionalities beyond the CQDs' intrinsic properties. Infrared detectors with polarization

Citation: Tang, X. Special Issue "Application of Advanced Quantum Dots Films in Optoelectronics". *Coatings* **2023**, *13*, 589. https://doi.org/10.3390/coatings13030589

Received: 3 March 2023
Accepted: 7 March 2023
Published: 9 March 2023

sensitivity could extend the information dimension of the detected signals and improve the target recognition ability. However, traditional infrared polarization detectors with epitaxial semiconductors usually suffer from a low extinction ratio, complexity in structure, and high cost. In the work by Zhao et al. [6], a simulation study of CQD infrared detectors with a monolithically integrated metal wire grid polarizer and an optical cavity was conducted. The polarization selectivity of HgTe CQDs with resonant-cavity-enhanced wire grid polarizers was studied in both shortwave and midwave infrared regions. Besides the high extinction ratio, the optical-cavity-enhanced wire grid polarizer could also significantly improve light absorption by a factor of 1.5, which leads to higher quantum efficiency and better spectral selectivity. With a meta-lens as a light concentrator, an over-20-times enhancement in light absorption can be achieved [7].

Besides infrared, X-ray detectors also demonstrate a wide range of applications. Halide perovskite has remarkable optoelectronic properties, such as high atomic number, large carrier mobility–lifetime product, high X-ray attenuation coefficient, and simple and low-cost synthesis process, and has gradually developed into a next-generation X-ray detection material. In the work by Tan et al., the fabrication methods of halide perovskite film and the development progress of halide-perovskite-based X-ray detectors were introduced [8].

Emerging infrared upconversion imaging devices can directly convert low-energy infrared photons into high-energy visible-light photons; thus, they hold promise in accomplishing pixel-less high-resolution infrared imaging at low cost. In the work by Rao et al., the recent advances and progress of infrared-to-visible upconversion devices were summarized [9]. To further improve the performance of infrared upconverters, one possible direction is to upgrade single-color upconverters to multicolor upconverters, which can correlate the color of emitted light with infrared light intensity, temperature, pressure, or biosignals. The key component, a multicolor light-emitting diode, was discussed by Ma et al. [10].

Funding: This work was funded by National Key R&D Program of China (2021YFA0717600), National Natural Science Foundation of China (NSFC No. 62035004). X.T. is sponsored by the Young Elite Scientists Sponsorship Program by CAST (No. YESS20200163).

Conflicts of Interest: The authors declare no conflict of interest.

References

1. Zhang, S.; Hu, Y.; Hao, Q. Advances of Sensitive Infrared Detectors with HgTe Colloidal Quantum Dots. *Coatings* **2020**, *10*, 760. [CrossRef]
2. Li, T.; Tang, X.; Chen, M. Room-Temperature Infrared Photodetectors with Zero-Dimensional and New Two-Dimensional Materials. *Coatings* **2022**, *12*, 609. [CrossRef]
3. Bi, C.; Liu, Y. CMOS-Compatible Optoelectronic Imagers. *Coatings* **2022**, *12*, 1609. [CrossRef]
4. Zhao, X.; Mu, G.; Tang, X.; Chen, M. Mid-IR Intraband Photodetectors with Colloidal Quantum Dots. *Coatings* **2022**, *12*, 467. [CrossRef]
5. Wen, C.; Zhao, X.; Mu, G.; Chen, M.; Tang, X. Simulation and Design of HgSe Colloidal Quantum-Dot Microspectrometers. *Coatings* **2022**, *12*, 888. [CrossRef]
6. Zhao, P.; Mu, G.; Chen, M.; Tang, X. Simulation of Resonant Cavity-Coupled Colloidal Quantum-Dot Detectors with Polarization Sensitivity. *Coatings* **2022**, *12*, 499. [CrossRef]
7. Ning, Y.; Zhang, S.; Hu, Y.; Hao, Q.; Tang, X. Simulation of Monolithically Integrated Meta-Lens with Colloidal Quantum Dot Infrared Detectors for Enhanced Absorption. *Coatings* **2020**, *10*, 1218. [CrossRef]
8. Tan, Y.; Mu, G.; Chen, M.; Tang, X. X-ray Detectors Based on Halide Perovskite Materials. *Coatings* **2023**, *13*, 211. [CrossRef]
9. Rao, T.; Chen, M.; Mu, G.; Tang, X. Infrared-to-Visible Upconversion Devices. *Coatings* **2022**, *12*, 456. [CrossRef]
10. Ma, S.; Qi, Y.; Mu, G.; Chen, M.; Tang, X. Multi-Color Light-Emitting Diodes. *Coatings* **2023**, *13*, 182. [CrossRef]

Review

Advances of Sensitive Infrared Detectors with HgTe Colloidal Quantum Dots

Shuo Zhang, Yao Hu and Qun Hao *

Beijing Key Laboratory for Precision Optoelectronic Measurement Instrument and Technology, School of Optics and Photonics, Beijing Institute of Technology, Beijing 100081, China; 3120195342@bit.edu.cn (S.Z.); huy08@bit.edu.cn (Y.H.)

* Correspondence: qhao@bit.edu.cn

Received: 8 July 2020; Accepted: 30 July 2020; Published: 4 August 2020

Abstract: The application of infrared detectors based on epitaxially grown semiconductors such as HgCdTe, InSb and InGaAs is limited by their high cost and difficulty in raising operating temperature. The development of infrared detectors depends on cheaper materials with high carrier mobility, tunable spectral response and compatibility with large-scale semiconductor processes. In recent years, the appearance of mercury telluride colloidal quantum dots (HgTe CQDs) provided a new choice for infrared detection and had attracted wide attention due to their excellent optical properties, solubility processability, mechanical flexibility and size-tunable absorption features. In this review, we summarized the recent progress of HgTe CQDs based infrared detectors, including synthesis, device physics, photodetection mechanism, multi-spectral imaging and focal plane array (FPA).

Keywords: HgTe Colloidal quantum dots; infrared; photoconductors; photovoltaic; multispectral imaging

---

1. Introduction

Infrared detectors are widely used in thermal imaging [1,2], material spectroscopy analysis [3–5], autonomous driving assistants [6], surveillance [7] and biological health monitoring [8–10]. Before 1940s, the dominant infrared detectors were thermal detector, such as thermocouple detector [11], bolometer [12] and pyroelectric detector [13], which mainly relied on the detection of incident power rather than the energy of the incoming photons. Despite that they can be operated at room temperature, thermal detectors suffer from slower response speed and lower detectivity compared with photon detectors. Driven by the demands for fast response and high sensitivity, photon detectors gradually occupied the dominant position of infrared detection technology. The first photon detector was a PbS detector with a response wavelength of 3 μm, and germanium-mercury alloy (Ge:Hg), lead selenide (PbSe), germanium-silicon alloy (Ge:Si), indium antimonide (InSb) and mercury cadmium telluride (HgCgTe) infrared detectors were successively developed. Infrared detectors and imaging systems have gone through "scan to image" and "single-color focal plane array (FPA)", and currently are moving towards large-format, multi-color 3rd and 4th generations [14]. Figure 1 showed the development roadmap of infrared detectors. The first generation of infrared detectors systems were relatively small refrigeration type photodetector unit or linear array, used with a complex two-dimensional scanning system to obtain images [15,16]. Compared with the first generation systems, the second generation systems had more pixels distributed on the two-dimensional FPA [17]. These pixels and readout integrated circuits (ROICs) constituted the infrared detection system, imaging by staring system electrically scanned. The third and fourth generation systems are being developed and are not well defined. In general, these systems are required to have large number of pixels, high frame rate, multi-color detection, integrated signal processing and other powerful functions.

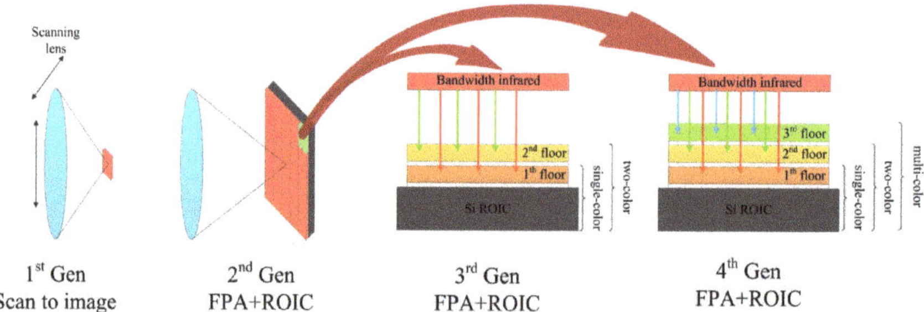

Figure 1. The development roadmap of infrared detectors. Through scan to image and single-color FPA towards two-color and multi-color.

Nowadays, thermal imaging relies primarily on bulk epitaxial semiconductor materials. In mid-wave infrared (MWIR, 3–5 μm) and long-wave infrared (LWIR, 8–12 μm), it is dominated by narrow bandgap semiconductor material such as InSb [18], HgCgTe [19,20], quantum-wells [21] and type-II superlattices [22,23]. However, these materials were typically prepared by molecular beam epitaxy or chemical vapor deposition, which leads to high fabrication and processing costs. The applications of epitaxial materials-based infrared detectors are, therefore, limited to defense and scientific research, and there is an urgent need for new infrared materials that combines fast response, high sensitivity and compatibility with CMOS semiconductor processes to promote civilian applications, such as driverless car, atmospheric monitoring, thermal imaging, agricultural planting, etc.

As an alternative to conventional bulk epitaxial semiconductors, colloidal quantum dots (CQDs) have attracted wide attention due to their unique advantageous features such as the ease of solution-processing, low-cost and large-scale synthesis and size-tunable optical absorption wavelength. Over the past decade, CQDs have been widely used in a variety of applications, including solar cells [24], spectrometers [25], phototransistors [26], FPA imagers [27], lasers [28], LEDs [29]. Among all the CQDs systems, mercury telluride (HgTe) CQDs have demonstrated the highest infrared spectral absorption tunability covering main important atmospheric windows including short-wave infrared (SWIR, 1.5–2.5 μm) [30–33], mid-wave infrared (MWIR, 3–5 μm) [34–42], long-wave infrared (LWIR, 8–12 μm) [43,44] and even the terahertz (THz) [45,46], appearing as promising candidates to substitute conventional semiconductors to achieve good detection performance with low cost. This review will be described from the following aspects, including: synthesis of infrared CQDs, infrared CQDs photodetectors, multispectral CQDs photodetectors and CQDs FPA.

2. Synthesis of Infrared CQDs

HgTe CQDs, a kind of zero-gap intrinsic semiconductor in the bulk, had a dominant position in the mid-infrared band of colloidal nanomaterials with interband optical transitions. In principle, the band gap can be adjusted in the full infrared range by controlling CQDs size. The synthesis [34,35] and optical properties [47] of HgTe CQDs with band-gaps tunable through the mid-infrared range of 3–5 μm (0.2 to 0.5 eV) were first reported by Guyot-Sionnest group in the University of Chicago. The typical synthesis of monodispersed HgTe CQD involved the reaction of telluride (Te) and mercury chloride ($HgCl_2$) in non-polar solvents [35]. In brief, Te was dissolved in trioctylphosphine (TOP) and $HgCl_2$ was dissolved in oleylamine, forming a clear, colorless or pale-yellow solution. To initiate the reaction, TOP:Te was rapidly injected in to $HgCl_2$ solution. After a period of reaction time, the reaction was quenched by adding dodecanethiol and the final particle size depended on the temperature and duration of the reaction, typically in the range of 60 to 130 °C and 1 to 40 min. Transmission electron microscopy (TEM) image of HgTe CQDs with shape of triangle or distorted parallelogram is shown in Figure 2a. Absorption and photodetection with well-defined edges, as well as narrow photoluminescence, were tunable across the near and mid-IR between 1.3 and 5 μm. The typical infrared absorption spectra of different sizes are shown in Figure 2b.

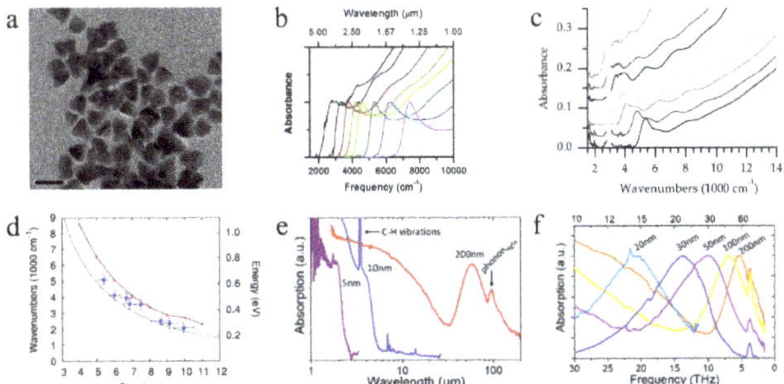

Figure 2. (**a**) Transmission electron microscopy image of HgTe CQOs, scale bar is 20 nm, adapted with permission from [35]; Copyright 2011 American Chemical Society. (**b**) The spectral absorption of HgTe CQDs with cut-off wavelength up to 5 μm, adapted with permission from [35]; Copyright 2011 American Chemical Society. (**c**) Absorption spectra of particles of different sizes show better spectral resolution, adapted with permission from [43]; Copyright 2014 American Chemical Society. (**d**) Blue dots: A functional relationship between the photoluminescence peak energy and the CQDs size measured in TEM images. Red line and crosses: calculation of exciton gap of truncated tetrahedral shape CQDs. Black dotted line: calculation of exciton gap versus diameter of spherical CQDs. Adapted with permission from [43]; Copyright 2014 American Chemical Society. (**e**) Absorption spectra of various sizes HgTe CQD, adapted with permission from [45]; Copyright 2018 American Chemical Society. (**f**) Absorption spectra of various sizes HgTe CQDs, presenting an intraband/plasmonic characteristic in the LWIR, very long-wavelength infrared range (VLWIR, 12–20 μm range) or THz range, adapted with permission from [45]; Copyright 2018 American Chemical Society.

In 2014, the Guyot-Sionnest group further improved the synthesis method of HgTe CQDs by making a slight modification by TOP:Te precursor dilution with oleylamine, which enabled it to display better spectral resolution, as shown in Figure 2c [43]. A more homogeneous mixture is formed by diluting Te precursors before appreciable growth occurs, and additional nucleation is prevented by sufficient cooling [43]. Through regrowing of particles by dropwise addition of $HgCl_2$ and TOP:Te dissolved in oleylamine at 110–120 °C, the absorption of LWIR was realized. The HgTe CQDs size dependence of the energy gaps was shown in Figure 2d. In 2017, a new method for synthesizing HgTe CQDs using more active tellurium sources and excess mercury precursors was reported by Guyot-Sionnest group, to produce more spherical and non-agglomerating tendency HgTe CQDs, which showed air-stable n-doping [48]. In 2018, Lhuillier Group in Sorbonne Université reported a synthesis method of HgTe CQD with an absorption cutoff wavelength covering the whole infrared range from 2 μm up to the THz range (65 μm), with corresponding HgTe CQD sizes ranging from 5 nm to above 200 nm [45], as shown in Figure 2e,f.

3. Infrared CQDs Photodetectors

CQDs have emerged as a powerful and versatile platform for building electronic and optoelectronic devices with their excellent tunable optoelectronic properties and low cost and large-scale synthesis. HgTe CQDs infrared photodetectors mainly include three categories: photoconductors, phototransistors and photovoltaics. The first reported HgTe CQDs photodetector is a photoconductor [34]. When the photon energy is greater than the band gap, the conductivity of the CQDs can be controlled by absorbing photons. The controls of surface ligand are a key step toward the design of photoconductive thin films. In most cases, CQDs synthesis requires the use of long-chain organic ligands, which can ensure the stable growth of the nanometer size of CQDs and enable the CQDs to overcome the van der Waals interparticle attraction for better stabilization in the form of colloids. However, long-chain ligands are not suitable for electronic applications. Initial long-chain ligands need to be exchanged by short-chain ligands to increase the inter-CQD coupling and obtain reasonable carrier mobility to increase the conductivity.

A more powerful infrared detector should be able to give a higher specific detectivity D^*. A high D^* means that the detector has a strong ability to detect weak signals. The D^* can be calculated as

$$D^* = \frac{(A\Delta f)^{1/2}}{NEP} = \frac{(A\Delta f)^{1/2}}{i_n}R \tag{1}$$

where A is detector area, Δf is the amplifier bandwidth, NEP is the noise equivalent power, i_n is the noise current and R is the responsivity. The responsivity is the ratio of the output voltage or current of the detector to the radiated power incident on the photosensitive region of the detector, can be calculated as:

$$R = \frac{I_{out}}{P} \tag{2}$$

$$R = \frac{V_{out}}{P} \tag{3}$$

where I_{out} is the output current of the detector, V_{out} is the output current voltage of the detector and P is the radiated power

The device architecture of the photoconductor is the simplest, which can be fabricated by depositing CQDs on a prefabricated interdigitated electrode [49], as shown in Figure 3a. Although the structure of the photoconductive device is simple, the performance of the device suffered from dark-current and $1/f$ noise [50], which limited the specific detectivity (D^*) of the mid-infrared HgTe CQDs photoconductors on the order of a few 10^9 Jones at 80 K [34].

The performance of CQDs photoconductors can be improved by using compact ligands, which allows higher mobility state. The hybrid ligand exchange and polar phase transfer technology were used to realize the improved mid-infrared photoconductors based on HgTe CQDs, which improved the mobility of carriers and accurately tuned the doping level of HgTe CQDs, so that the D^* of detector was increased to 4.5×10^{10} Jones at 80 K and 500 Hz, and a 5 μm cutoff wavelength [42].

Similar to photoconductors, the amount of photocurrent in a phototransistor varies according to the intensity of light, whereas phototransistors combined a biased high-mobility channel such as graphene and MoS_2 to achieve photoconductivity gain. The photoconductivity gain represents the ratio of the external photocurrent formed by the photo-generated carriers under bias voltage to the internal photocurrent formed by the photoelectron. Hybrid device architectures such as grapheme-CQDs [26,51], MoS_2-CQDs [52,53] and grapheme-MoS_2 [54] of phototransistors have been reported, demonstrating high photoconductivity gains and ultra-high optical sensitivities. However, the photoconductive gain amplifies not only the photocurrent but also any noise due to thermal excitation or recombination. The structure schematic diagram of a MoS_2-CQDs hybrid device is shown as Figure 3b.

Compared with HgTe CQDs photoconductors, the photovoltaic devices improved device speed and sensitivity, and theoretically avoided $1/f$ noise and dark current. In 2015, *Guyot-Sionnest* group developed the first background-limited infrared photodetection (BLIP) HgTe CQD mid-wave infrared detector, which achieved D^* of 4.2×10^{10} Jones, microseconds response times and covered all the infrared spectrum regions with a cut-off at 5.25 μm at 90 K [38]. Figure 3c give an illustration of the first BLIP HgTe CQDs photovoltaic detector and the function of current of the detector with the applied bias, at different ambient temperatures. In 2018, Guyot-Sionnest group further improved the sensitivity, efficiency, and response speed of the photovoltaic MWIR detectors through a new doping method [40]. A solid-state cation exchange method was introduced to achieve heavily p-doped HgTe CQDs by using Ag_2Te nanocrystals. The specific doping method is to spin-coating Ag_2Te nanocrystals onto HgTe CQDs and expose them to $HgCl_2$. Ag_2Te nanocrystals liberate Ag^+ ions and then form $AgCl_2$ and immobilize them on the surface of HgTe CQDs. This doping scheme increases the D* of the device by one or two orders of magnitude [40]. Schematic and cross-sectional Scanning electron microscopy image of the HgTe CQDs photodetectors are shown in Figure 3d,e [40]. The MWIR absorption in the typical thickness of ~400 nm HgTe film in the device is ~18% and the corresponding internal quantum efficiency (IQE) was estimated to be as high as 90% [40]. To further improve device performance, some optical structures have been reported, such as optical nano-antennas [39], plasmonic nano-disk array [41] and gold nanorods [55], which increased the D^* by two or three times. Figure 3f showed the schematic of the HgTe CQDs midwave infrared detector with enhanced light absorption by plasmonic nano-disks [41]. Mechanical flexibility and easy integration with flexible substrates without lattice matching is another advantage of CQDs over traditional semiconductor materials. In 2019, Tang et al. fabricated the first flexible HgTe CQDs photovoltaic detector on a polyimide substrate and integrated a Fabry-Perot resonator cavity to further improved detectivity [56]. The Fabry-Perot resonant cavity is composed of the reflection layer at the bottom, the gold semi-reflector at the top and the CQDs, ITO and optical spacer sandwiched in the middle [56]. The infrared light illuminated from the top bounced back and forth between the reflective layer at the bottom and the semi-reflective layer at the top, enhanced the infrared light at specific wavelength. The schematic of the flexible device is shown in Figure 3g. Besides enhancing absorption, optical structures can also be used for better spectral resolution. By combining HgTe CQD photovoltaic sensors with distributed Bragg mirror filter array in a single device, Tang et al, reported a CQDs hyperspectral sensor [57], which realized acquisition of hyperspectral data with 64 narrowband channels and full-width at half-maxima down to 30 cm^{-1}, as shown in Figure 3h.

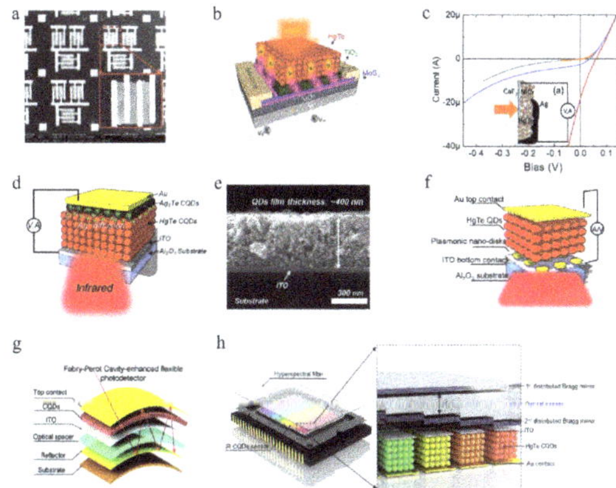

Figure 3. (**a**) Scanning electron microscopy image of a photoconductor by depositing CQDs on a prefabricated interdigitated electrode, adapted with permission from [49]; Copyright 2016 American Chemical Society. (**b**) Schematic of the fast and sensitive HgTe CQDs photodetectors: The structure schematic diagram of a MoS_2-CQDS hybrid device, adapted with permission from [52]; Copyright 2017 WILEY-VCH Verlag GmbH & Co. KGaA, Weinheim. (**c**) Illustration of the first BLIP HgTe CQDs photovoltaic detector. The function of current of the photoelectric device with the applied bias, at 90 K in the dark (black line), receiving environment 295 K radiation (blue line) and receiving 873 K blackbody lighting (red line). Reprinted with permission from [38]; Copyright 2015 AIP Publishing LLC. (**d**) Schematic of the fast and sensitive HgTe CQDs photodetectors, adapted with permission from [40]; Copyright 2018 American Chemical Society. (**e**) Cross-sectional Scanning electron microscopy image of the fast and sensitive HgTe CQDs photodetectors, adapted with permission from [40]; Copyright 2018 American Chemical Society. (**f**) Schematic of HgTe CQDs mid-wave infrared detector with enhanced light absorption by plasmonic disks, adapted with permission from [41]; Copyright 2018 American Chemical Society. (**g**) Schematic of a flexible HgTe CQDs photodetector with enhanced light absorption through a Fabry-Perot cavity, adapted with permission from [56]; Copyright 2019 WILEY-VCH Verlag GmbH & Co. KGaA, Weinheim. (**h**) CQDs hyperspectral detector was composed of HgTe CQDs detector and hyperspectral filter, adapted with permission from [57]; Copyright 2019 WILEY-VCH Verlag GmbH & Co. KGaA, Weinheim.

In order to better reflect the development of infrared CQDs photodetectors, Table 1 shows the performance comparison between HgTe CQDs photodetectors over the past decade.

Table 1. Performance comparison between HgTe CQDs photodetectors.

Device Structure Type	Year	Spectral Range (μm)	*R* (A/W)	EQE (%)	*D** (Jones)	Response Time	Reference
HgTe CQDs photoconductors	2011	<5	0.25	10	2×10^9	NA	[34]
HgTe CQDs photoconductors	2014	<12	3×10^{-4}	NA	6.5×10^6	<5 μs	[43]
HgTe CQDs photoconductors	2019	<5	0.2	30	4.5×10^{10}	NA	[42]
HgTe/AsS_3 phototransistors	2013	<3.5	5×10^{-3}	NA	3.5×10^{10}	NA	[37]
HgTe CQDs phototransistors	2017	<2	0.4	NA	2×10^{10}	NA	[31]
MoS_2-HgTe CQDs Hybrid phototransistors	2017	<2	10^6	NA	10^{12}	4 ms	[52]
HgTe CQDs photodiodes	2015	3–5	8×10^{-2}	2.5	4.2×10^{10}	0.7 μs	[38]
HgTe CQDs photodiodes	2018	<4.8	0.38	17	1.2×10^{11}	<1 μs	[40]
Plasmon ResonanceEnhanced HgTe CQDs photodiodes	2018	<4.5	1.62	45	4×10^{11}	<1 μs	[41]
Flexible HgTe CQDs photodiodes	2019	<2.2	0.5	30	7.5×10^{10}	260 ns	[56]
Hyperspectral HgTe CQDs photodiodes	2019	1.53–2.08	0.2	11	$>10^{10}$	120 ns	[57]
Dual-band HgTe CQDs photodiodes	2019	<2.5 and 3–5	0.3 and 0.15	NA	10^{11} and 3×10^{10}	<2.5 μs	[58]
HgTe CQDs photodiodes	2020	2–3	1	30	10^{11}	NA	[33]

4. Multispectral CQDs Photodetectors

According to the development history of infrared detectors and systems, the third and fourth generation of infrared imaging requires simultaneous detection of two or more wavelengths in photoresponse. Multispectral detection provides better target identification and allows accurate determination of the infrared signature of the target.

By vertically stacking one SWIR and one MWIR HgTe CQD photodiode into an n-p-n structure, Tang et al. fabricated the first CQDs two-terminal dual-band infrared detector in 2019 to switch the device's spectral response between short-wave and mid-wave infrared by controlling the bias polarity and magnitude [58]. Compared to a three-terminal configuration using a common ground contact [2], this two-terminal structure increased the optical fill factor. The architecture and cross-sectional Scanning electron microscopy image of the two-terminal dual-band device were shown in Figure 4a,b with Bi_2Se_3 and Ag_2Te as the n and p layers. Figure 4c showed the simplified energy diagram of the device at zero bias, indicating the n-p-n structure of the device. When the bias voltage of the detector gradually changes from −300 to +500 mV, the spectral response of the device gradually switched from SWIR to MWIR, as shown in Figure 4d. The two-terminal dual-band detector can produce SWIR image and MWIR image of object respectively, and can also generate the combined two-color image. The obvious two-color image can provide more information than the single image. The SWIR, MWIR and two-color composite images of cold and hot water were shown in Figure 4e. In addition to imaging, the two-terminal dual-band detector can be used to determine the temperature of object by the SWIR/MWIR signal ratio, which is more reliable than single-band detection.

Although the two-terminal dual-band detector with architecture of vertically stacked two different-sized HgTe CQD photodiodes [58] has superior performance, it is limited to two-color functionality. It is difficult to achieve more spectral channels with vertical stacks, which requires more complex band engineering. Although the spatial resolution and the optical fill factor would be reduced, pixelating CQDs of different sizes in different areas of a chip would seem to be a better solution for multicolor detection. Therefore, the key to achieving the transition from monochromatic detectors to multicolor detectors with CQDs is to develop CQDs patterning methods to selectively deposit different size CQDs with well-defined geometry in different areas. Several CQDs patterning techniques have been demonstrated to optimize the resolution, throughput and fidelity, including lithographical lift-off [59,60], contact printing [61], transfer printing [49,62,63], inkjet printing [30,64,65], screen printing [66] and laser writing [67].

A simple and efficient Poly(methyl methacrylate) (PMMA)-assisted transfer technique was reported for the scalable fabrication CQD-based multicolor photodetector with multi-pixels in response to different spectral ranges of infrared by using patterned HgTe CQD films [49], as shown in Figure 5a,b. However, during CQDs patterning, photoresist residue, developer, heating, or plasma etching may change the properties of CQDs.

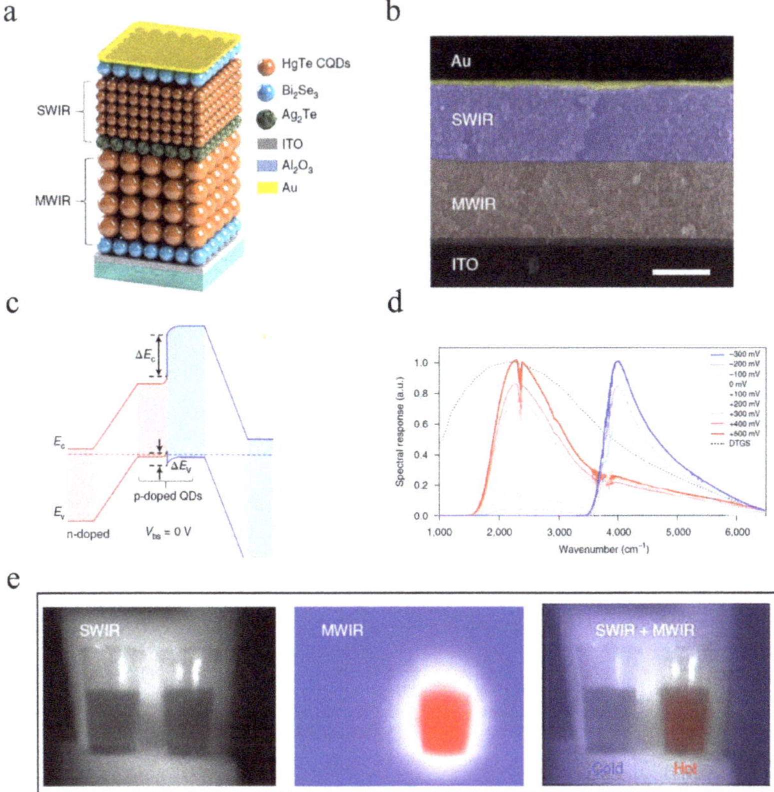

Figure 4. (**a**) Architecture of the two-terminal dual-band detector. (**b**) Cross-sectional Scanning electron microscopy image of the detector. Scale bar: 300 nm. (**c**) Energy diagram of the two-terminal dual band detector at zero bias. (**d**) The spectral response of the detector varies with bias voltage from −300 to +500 mV. (**e**) Cold and hot water's SWIR, MWIR and two-color composite images. Adapted with permission from [58]; Copyright 2019 Nature Publishing Group.

Wang et al. reported a photoresists-free technique for direct optical patterning of CQDs, called direct optical lithography of functional inorganic nanomaterials (DOLFIN), and designed a series of photosensitive ligands to build an extensive library of functional photosensitive inorganic inks [68,69]. Under controlled light exposure, the compact photosensitive surface ligands of CQDs undergo chemical transformations that alter the solubility of CQDs in the developer. The resolution of the CQDs patterning produced by this method is comparable with that of traditional photolithography, without compromising the electronic and optical properties. Using similar patterning techniques, Kim et al. demonstrated a low-temperature fabricated two-dimensionally pixelized full-color photodetector by using monolithic integration of various-sized CQDs and amorphous indium-gallium-zinc-oxide semiconductors [70]. As a parallel technology, DOLFIN has various advantages, but specific ligands are required. Further verification is needed to determine whether the influence of DOLFIN on the synthesis of HgTe CODs and precise doping in HgTe CQDs photovoltaic devices, as well as whether DOLFIN can be applied to HgTe CQDs multi-spectral infrared detector. Looking at the various CQDs patterning techniques, inkjet printing appears to be the most promising for fabrication HgTe CQDs multi-spectral infrared detector. The key is to develop a fast and reliable parallel inkjet printing method.

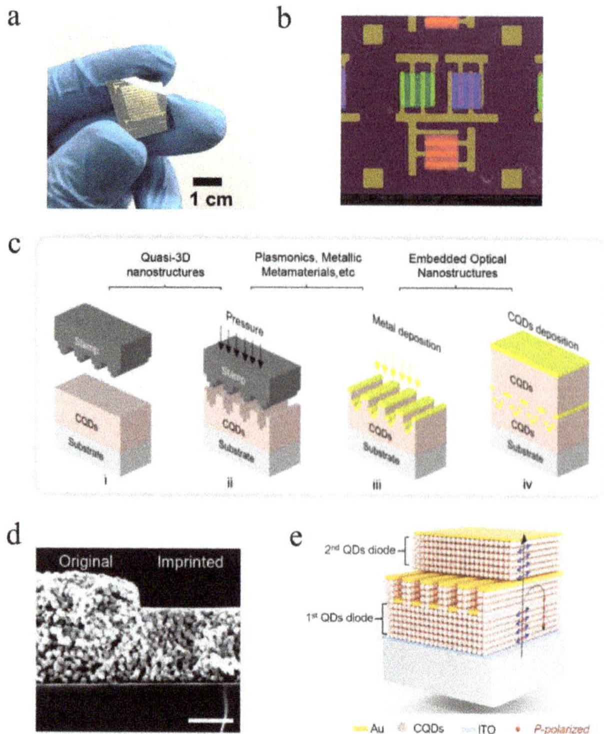

Figure 5. (**a**) A photo of a three-pixel photodetectors, adapted with permission from [49]; Copyright 2016 American Chemical Society. (**b**) False-colored Scanning electron microscopy image of the three-pixel photodetectors, adapted with permission from [49]; Copyright 2016 American Chemical Society. (**c**) Direct imprinting technique process, adapted with permission from [71]; Copyright 2020 WILEY-VCH Verlag GmbH & Co. KGaA, Weinheim. (**d**) Cross-sectional scanning electron microscopy image of imprinted CQDs film. Scale bar is 100 nm, adapted with permission from [71]; Copyright 2020 WILEY-VCH Verlag GmbH & Co. KGaA, Weinheim. (**e**) Illustration of the two-terminal dual-channel infrared polarization detector with the imprinting technology, adapted with permission from [71]; Copyright 2020 WILEY-VCH Verlag GmbH & Co. KGaA, Weinheim.

Besides 2D CQDs patterning techniques, Tang et al. reported a direct imprinting technique for producing high resolution CQDs quasi-3D optical nanostructures without degrading the key properties of materials [71]. The direct imprinting technique process was shown in Figure 5c. Firstly, nano-structured silicon stamps were prepared by using traditional semiconductor technology; the nanostructures were then transferred to the CQDs film by contacting the CQDs with the stamp and applying pressure; last, by metal deposition and capping another layer of CQDs, the high-resolution nanophotonic structure is embedded into the device. A cross-sectional scanning electron microscopy image of imprinted CQDs film is shown in Figure 5d. The imprinting pattern has clear and sharp edges, no large cracks or pile-ups of CQDs, and the surface roughness of the imprinting area is reduced, which is beneficial and necessary for high-precision patterning. With the imprinting technology, a two-terminal dual-channel infrared polarization detector was designed to detect both polarized and unpolarized light, as shown in Figure 5e. A bilayer wire-grid polarizer embedded in CQDs can be used as optical polarization filter and middle electrode for signal readout.

5. CQDs Focal Plane Array

In conventional HgCdTe or InSb FPA infrared detectors, the active layer is prepared by high-temperature epitaxial growth, which is a very expensive technique, and must be coupled to ROIC by indium bumps, which further increases the cost. When building an FPA, the pixel size should be ideally just above the target wavelength to improve image quality. The pixel size of HgCdTe or InSb FPA infrared detectors is actually limited by technology and more specifically by the indium bump hybridization to the ROICs.

As an interesting candidate, semiconductor CQD can produce infrared detectors by simple drop-casting or spin-coating on a silicon ROICs, without molecular beam epitaxy and lattice-match with the substrate. This greatly reduces the cost of the monolithic FPA infrared detectors, which is an unmatched advantage over traditional semiconductors. In 2016, Guyot-Sionnest group reported the first CQDs FPA infrared detector for mid-infrared imaging by coupling the HgTe CQDs with a silicon ROICs [72]. A study by Guyot-Sionnest Group demonstrated a simple fabrication method for a low-cost HgTe CQDs FPA thermal camera with noise equivalent difference (NEDT) of 1.02 mK at 5 μm and achieved images at 120 frames per second using the CQDs FPA infrared detector [73]. CQDs have been demonstrated to be a promising material for infrared imaging due to its excellent optoelectronic properties, low cost, easy synthesis, and mechanical flexibility.

6. Conclusions and Outlook

After more than a decade of intensive research, HgTe CQDs can be considered as a possible alternative to conventional semiconductor infrared sensing. The main achievements have been demonstrated include full tunability of the absorption over the infrared range, the fast detectors with background-limited infrared photodetection, flexible and curved infrared detector with high D^* and short response time, and the CQDs-based FPA in both SWIR and MWIR range.

CQDs bring both opportunities and challenges to the development of infrared detectors. Although synthetic improvements in CQDs are still possible, surface chemistry, film processing, and doping control require further research to achieve higher detection capabilities at higher temperatures. On the other hand, the CQDs infrared detectors reported at the present stage mainly focus on the SWIR and MWIR, while there are few reports on the LWIR and its performance is poor compared with that of commercial detectors. Therefore, the CQDs LWIR device is an urgent direction to be explored. At the same time, an efficient, accurate and large-scale CQDs patterning technology is needed to meet the requirements of FPA and multi-spectral imaging. For the future application of civilian and military infrared detectors, HgTe CQDs will meet many serious challenges. Most of the reported HgTe CQDs infrared detectors were single-pixel devices with less FPA and worse performance than commercial infrared detectors. For many systems such as night vision, wide range monitoring and astronomical applications that require large FPA and high pixel densities, an efficient CQDs pixelation method is required. In addition, the performance stability of HgTe CQDs infrared detector is a challenge in some extreme environments such as high temperature, high pressure and space environment.

Author Contributions: Investigation, S.Z.; writing—original draft preparation, S.Z.; writing—review and editing, Y.H. and Q.H.; project administration, Y.H.; funding acquisition, Q.H. All authors have read and agreed to the published version of the manuscript.

Funding: The work was supported by the National Natural Science Foundation of China (NSFC) (51735002 and 61875012).

Conflicts of Interest: The authors declare no conflict of interest.

References

1. Rogalski, A. Toward third generation HgCdTe infrared detectors. *J. Alloy. Compd.* **2004**, *371*, 53–57. [CrossRef]
2. Rogalski, A.; Antoszewski, J.; Faraone, L. Third-generation infrared photodetector arrays. *J. Appl. Phys.* **2009**, *105*, 091101. [CrossRef]

3. Stouwdam, J.W.; Veggel, F.C.J.M.V. Near-infrared emission of redispersible Er^{3+}, Nd^{3+}, and Ho^{3+} doped LaF_3 nanoparticles. *Nano Lett.* **2002**, *2*, 733–737. [CrossRef]
4. Schmitt, J.; Flemming, H.C. FTIR-spectroscopy in microbial and material analysis. *Int. Biodeterior. Biodegrad.* **1998**, *41*, 1–11. [CrossRef]
5. Madejová, J. FTIR techniques in clay mineral studies. *Vib. Spectrosc.* **2003**, *31*, 1–10. [CrossRef]
6. González, A.; Fang, Z.; Socarras, Y.; Serrat, J.; Vázquez, D.; Xu, J.; López, A.M. Pedestrian detection at day/night time with visible and FIR cameras: A comparison. *Sensors* **2016**, *16*, 820. [CrossRef]
7. Briz1, S.; de Castro, A.J.; Aranda, J.M.; Mele´ndez, J.; Lo´pez, F. Reduction of false alarm rate in automatic forest fire infrared surveillance systems. *Remote Sens. Environ.* **2003**, *86*, 19–29. [CrossRef]
8. Soref, R. Mid-infrared photonics in silicon and germanium. *Nat. Photonics* **2001**, *4*, 495–497. [CrossRef]
9. Miller, L.M.; Smity, G.D.; Carr, G.L. Synchrotron-based biological microspectroscopy: From the mid-infrared through the far-infrared regimes. *J. Biol. Phys.* **2003**, *29*, 219–230. [CrossRef]
10. Yokota, T.; Zalar, P.; Kaltenbrunner, M.; Jinno, H.; Matsuhisa, N.; Kitanosako, H.; Tachibana, Y.; Yukita, W.; Koizumi, M.; Someya, T. Ultraflexible organic photonic skin. *Sci. Adv.* **2016**, *2*, e1501856. [CrossRef]
11. Nelms, N.; Dowson, J. Goldblack coating for thermal infrared detectors. *Sens. Actuators A* **2005**, *120*, 403–407. [CrossRef]
12. Langley, S.P. The Bolometer. *Nature* **1881**, *25*, 14–16.
13. Chynoweth, A.G. Dynamic method for measuring the pyroelectric effect with special reference to barium titanate. *J. Appl. Phys.* **1956**, *27*, 78. [CrossRef]
14. Martyniuk, P.; Antoszewski, J.; Martyniuk, M.; Faraone, L.; Rogalski, A. New concepts in infrared photodetector designs. *Appl. Phys. Rev.* **2014**, *1*, 41102. [CrossRef]
15. Elliott, C.T.; Day, D.; Wilson, D.J. An integrating detector for serial scan thermal imaging. *Infrared Phys.* **1982**, *22*, 31–42. [CrossRef]
16. Blackburn, A.; Blackman, M.V.; Charlton, D.E.; Dunn, W.A.E.; Jenner, M.D.; Oliver, K.J.; Wotherspoon, J.T.M. The practical realisation and performance of sprite detectors. *Infrared Phys.* **1982**, *22*, 57–64. [CrossRef]
17. Scribner, D.A.; Kruer, M.R.; Killiany, J.M. Infrared focal plane array technology. *Proc. IEEE* **1991**, *79*, 66–85. [CrossRef]
18. Camargo, E.G.; Ueno, K.; Morishita, T.; Sato, M.; Endo, H.; Kurihara, M.; Ishibashi, K.; Kuze, N. High-sensitivity temperature measurement with miniaturized InSb mid-IR sensor. *IEEE Sens. J.* **2007**, *7*, 1335–1339. [CrossRef]
19. Figgemeier, H.; Ames, C.; Beetz, J.; Breiter, R.; Eich, D.; Hanna, S.; Mahlein, K.M.; Schallenberg, T.; Sieck, A.; Wenisch, J. High-performance SWIR/MWIR and MWIR/MWIR bispectral MCT detectors by AIM. *Infrared Technol. Appl. Xliv.* **2018**, *10624*, 106240S. [CrossRef]
20. Cervera, C.; Baier, N.; Gravrand, O.; Mollard, L.; Lobre, C.; Destefanis, G.; Zanatta, J.P.; Boulade, O.; Moreau, V. Low-dark current p-on-n MCT detector in long and very long-wavelength infrared. *Infrared Technol. Appl. Xli.* **2015**, *9451*, 945129. [CrossRef]
21. Sarusi, G. QWIP or other alternative for third generation infrared systems. *Infrared Phys. Technol.* **2003**, *44*, 439–444. [CrossRef]
22. Haddadi, A.; Dehzangi, A.; Chevallier, R.; Adhikary, S.; Razeghi, M. Bias-selectable nBn dual-band long-/very long-wavelength infrared photodetectors based on InAs/$InAs_{1-x}Sb_x$/$AlAs_{1-x}Sb_x$ type-II superlattices. *Sci. Rep.* **2017**, *7*, 3379. [CrossRef]
23. Haddadi, A.; Chevallier, R.; Chen, G.; Hoang, A.M.; Razeghi, M. Bias-selectable dual-band mid-/long-wavelength infrared photodetectors based on InAs/$InAs_{1-x}Sb_x$ type-II superlattices. *Appl. Phys. Lett.* **2015**, *106*, 011104. [CrossRef]
24. Wang, X.; Koleilat, G.I.; Tang, J.; Liu, H.; Kramer, I.J.; Debnath, R.; Brzozowski, L.; Barkhouse, D.A.R.; Levina, L.; Hoogland, S.; et al. Tandem colloidal quantum dot solar cells employing a graded recombination layer. *Nat. Photon.* **2011**, *5*, 480–484. [CrossRef]
25. Bao, J.; Bawendi, M.G. A colloidal quantum dot spectrometer. *Nature* **2015**, *523*, 67–70. [CrossRef]
26. Konstantatos, G.; Badioli, M.; Gaudreau, L.; Osmond, J.; Bernechea, M.; de Arquer, F.P.G.; Gatti, F.; Koppens, F.H.L. Hybrid graphene–quantum dot phototransistors with ultrahigh gain. *Nat. Nanotechnol.* **2012**, *7*, 363–368. [CrossRef]

27. Goossens, S.; Navickaite, G.; Monasterio, C.; Gupta, S.; Piqueras, J.J.; Pérez, R.; Burwell, G.; Nikitskiy, I.; Lasanta, T.; Galán, T.; et al. Broadband image sensor array based on graphene–CMOS integration. *Nat. Photon.* **2017**, *11*, 366–371. [CrossRef]
28. Wu, K.; Park, Y.; Lim, J.; Klimov, V.I. Towards zero-threshold optical gain using charged semiconductor quantum dots. *Nat. Nanotechnol.* **2017**, *12*, 1140–1147. [CrossRef]
29. Yuan, F.; Yuan, T.; Sui, L.; Wang, Z.; Xi, Z.; Li, Y.; Li, X.; Fan, L.; Tan, Z.; Chen, A.; et al. Engineering triangular carbon quantum dots with unprecedented narrow bandwidth emission for multicolored LEDs. *Nat. Commun.* **2018**, *9*, 2249. [CrossRef]
30. Böberl, M.; Kovalenko, M.V.; Gamerith, S.; List, E.J.W.; Heiss, W. Inkjet-printed nanocrystal photodetectors operating up to 3 μm wavelengths. *Adv. Mater.* **2007**, *19*, 3574–3578. [CrossRef]
31. Chen, M.; Lu, H.; Abdelazim, N.M.; Zhu, Y.; Wang, Z.; Ren, W.; Kershaw, S.V.; Rogach, A.L.; Zhao, N. Mercury telluride quantum dot based phototransistor enabling high-sensitivity room-temperature photodetection at 2000 nm. *ACS Nano* **2017**, *11*, 5614–5622. [CrossRef] [PubMed]
32. Jagtap, A.; Goubet, N.; Livache, C.; Chu, A.; Martinez, B.; Gréboval, C.; Qu, J.; Dandeu, E.; Becerra, L.; Witkowski, N.; et al. Short wave infrared devices based on HgTe nanocrystals with air stable performances. *J. Phys. Chem. C* **2018**, *122*, 14979–14985. [CrossRef]
33. Ackerman, M.M.; Chen, M.; Guyot-Sionnest, P. HgTe colloidal quantum dot photodiodes for extended short-wave infrared detection. *Appl. Phys. Lett.* **2020**, *116*, 083502. [CrossRef]
34. Keuleyan, S.; Lhuillier, E.; Brajuskovic, V.; Guyot-Sionnest, P. Mid-infrared HgTe colloidal quantum dot photodetectors. *Nat. Photonics* **2011**, *5*, 489–493. [CrossRef]
35. Keuleyan, S.; Lhuillier, E.; Guyot-Sionnest, P. Synthesis of colloidal HgTe quantum dots for narrow mid-IR emission and detection. *J. Am. Chem. Soc.* **2011**, *133*, 16422–16424. [CrossRef] [PubMed]
36. Lhuillier, E.; Keuleyan, S.; Guyot-Sionnest, P. Colloidal quantum dots for mid-IR applications. *Infrared Phys. Technol.* **2013**, *59*, 133–136. [CrossRef]
37. Lhuillier, E.; Keuleyan, S.; Zolotavin, P.; Guyot-Sionnest, P. Mid-infrared HgTe/As_2S_3 field effect transistors and photodetectors. *Adv. Mater.* **2013**, *25*, 137–141. [CrossRef]
38. Guyot-Sionnest, P.; Roberts, J.A. Background limited mid-infrared photodetection with photovoltaic HgTe colloidal quantum dots. *Appl. Phys. Lett.* **2015**, *107*, 253104. [CrossRef]
39. Yifat, Y.; Ackerman, M.; Guyot-Sionnest, P. Mid-IR colloidal quantum dot detectors enhanced by optical nano-antennas. *Appl. Phys. Lett.* **2017**, *110*, 41106. [CrossRef]
40. Ackerman, M.M.; Tang, X.; Guyot-Sionnest, P. Fast and sensitive colloidal quantum dot mid-wave infrared photodetectors. *ACS Nano* **2018**, *12*, 7264–7271. [CrossRef]
41. Tang, X.; Ackerman, M.M.; Guyot-Sionnest, P. Thermal Imaging with Plasmon Resonance Enhanced HgTe Colloidal Quantum Dot Photovoltaic Devices. *ACS Nano* **2018**, *12*, 7362–7370. [CrossRef] [PubMed]
42. Chen, M.; Lan, X.; Tang, X.; Wang, Y.; Hudson, M.H.; Talapin, D.V.; Guyot-Sionnest, P. High carrier mobility in HgTe quantum dot solids improves mid-IR photodetectors. *ACS Photonics.* **2019**, *6*, 2358–2365. [CrossRef]
43. Keuleyan, S.E.; Guyot-Sionnest, P.; Delerue, C.; Allan, G. Mercury telluride colloidal quantum dots: Electronic structure, size-dependent spectra, and photocurrent detection up to 12 μm. *ACS Nano* **2014**, *8*, 8676–8682. [CrossRef] [PubMed]
44. Tang, X.; Wu, G.F.; Lai, K.W.C. Plasmon resonance enhanced colloidal HgSe quantum dot filterless narrowband photodetectors for mid-wave infrared. *J. Mater. Chem. C* **2017**, *5*, 362–369. [CrossRef]
45. Goubet, N.; Jagtap, A.; Livache, C.; Martinez, B.; Portalès, H.; Xu, X.Z.; Lobo, R.P.S.M.; Dubertret, B.; Lhuillier, E. Terahertz HgTe nanocrystals: Beyond confinement. *J. Am. Chem. Soc.* **2018**, *140*, 5033–5036. [CrossRef]
46. Lhuillier, E.; Scarafagio, M.; Hease, P.; Nadal, B.; Aubin, H.; Xu, X.Z.; Lequeux, N.; Patriarche, G.; Ithurria, S.; Dubertret, B. Infrared photodetection based on colloidal quantum-dot films with high mobility and optical absorption up to THz. *Nano Lett.* **2016**, *16*, 1282–1286. [CrossRef]
47. Lhuillier, E.; Keuleyan, S.; Guyot-Sionnest, P. Optical properties of HgTe colloidal quantum dots. *Nanotechnology* **2012**, *23*, 175705. [CrossRef]
48. Shen, G.; Chen, M.; Guyot-Sionnest, P. Synthesis of nonaggregating HgTe colloidal quantum dots and the emergence of air-stable n-doping. *J. Phys. Chem. Lett.* **2017**, *8*, 2224–2228. [CrossRef]
49. Tang, X.; Tang, X.; Lai, K.W.C. Scalable fabrication of infrared detectors with multispectral photoresponse based on patterned colloidal quantum dot films. *ACS Photonics* **2016**, *3*, 2396–2404. [CrossRef]

50. Liu, H.; Lhuillier, E.; Guyot-Sionnest, P. 1/f noise in semiconductor and metal nanocrystal solids. *J. Appl. Phys.* **2014**, *115*, 154309. [CrossRef]
51. Nikitskiy, I.; Goossens, S.; Kufer, D.; Lasanta, T.; Navickaite, G.; Koppens, F.H.L.; Konstantatos, G. Integrating an electrically active colloidal quantum dot photodiode with a graphene phototransistor. *Nat. Commun.* **2016**, *7*, 11954. [CrossRef] [PubMed]
52. Huo, N.; Gupta, S.; Konstantatos, G. MoS_2-HgTe Quantum dot hybrid photodetectors beyond 2 μm. *Adv. Mater.* **2017**, *29*, 1606576. [CrossRef] [PubMed]
53. Özdemir, O.; Ramiro, I.; Gupta, S.; Konstantatos, G. High sensitivity hybrid PbS CQD-TMDC photodetectors up to 2 μm. *ACS Photonics* **2019**, *6*, 2381–2386. [CrossRef]
54. Xu, H.; Wu, J.; Feng, Q.; Mao, N.; Wang, C.; Zhang, J. High responsivity and gate tunable graphene-MoS_2 hybrid phototransistor. *Small* **2014**, *10*, 2300–2306. [CrossRef] [PubMed]
55. Chen, M.; Shao, L.; Kershaw, S.V.; Yu, H.; Wang, J.; Rogach, A.L.; Zhao, N. Photocurrent enhancement of HgTe quantum dot photodiodes by plasmonic gold nanorod structures. *ACS Nano* **2014**, *8*, 8208–8216. [CrossRef] [PubMed]
56. Tang, X.; Ackerman, M.M.; Shen, G.; Guyot-Sionnest, P. Towards infrared electronic eyes: Flexible colloidal quantum dot photovoltaic detectors enhanced by resonant cavity. *Small* **2019**, *15*, 1804920. [CrossRef]
57. Tang, X.; Ackerman, M.M.; Guyot-Sionnest, P. Acquisition of hyperspectral data with colloidal quantum dots. *Laser Photonics Rev.* **2019**, *13*, 1900165. [CrossRef]
58. Tang, X.; Ackerman, M.M.; Chen, M.; Guyot-Sionnest, P. Dual-band infrared imaging using stacked colloidal quantum dot photodiodes. *Nat. Photonics* **2019**, *13*, 277–282. [CrossRef]
59. Choi, J.; Wang, H.; Oh, S.J.; Paik, T.; Jo, P.S.; Sung, J.; Ye, X.; Zhao, T.; Diroll, B.T.; Murray, C.B.; et al. Exploiting the colloidal nanocrystal library to construct electronic devices. *Science* **2016**, *352*, 205–208. [CrossRef]
60. Tang, X.; Chen, M.; Kamath, A.; Ackerman, M.M.; Guyot-Sionnest, P. Colloidal quantum-dots/graphene/silicon dual-channel detection of visible light and short-wave infrared. *ACS Photonics* **2020**, *7*, 1117–1121. [CrossRef]
61. Kim, L.; Anikeeva, P.O.; Coe-Sullivan, S.A.; Steckel, J.S.; Bawendi, M.G.; Bulovic, V. Contact printing of quantum dot light-emitting devices. *Nano Lett.* **2008**, *8*, 4513–4517. [CrossRef]
62. Kim, T.; Cho, K.; Lee, E.K.; Lee, S.J.; Chae, J.; Kim, J.W.; Kim, D.H.; Kwon, J.; Amaratunga, G.; Lee, S.Y.; et al. Full-colour quantum dot displays fabricated by transfer printing. *Nat. Photonics* **2011**, *5*, 176–182. [CrossRef]
63. Choi, M.K.; Yang, J.; Kang, K.; Kim, D.C.; Choi, C.; Park, C.; Kim, S.J.; Chae, S.I.; Kim, T.; Kim, J.H.; et al. Wearable red-green-blue quantum dot light-emitting diode array using high- resolution intaglio transfer printing. *Nat. Commun.* **2015**, *6*, 7149. [CrossRef] [PubMed]
64. Haverinen, H.M.; Myllylä, R.A.; Jabbour, G.E. Inkjet printing of light emitting quantum dots. *Appl. Phys. Lett.* **2009**, *94*, 73108. [CrossRef]
65. Wood, V.; Panzer, M.J.; Chen, J.; Bradley, M.S.; Halpert, J.E.; Bawendi, M.G.; Bulovic, V. Inkjet-printed quantum dot-polymer composites for full-color AC-driven displays. *Adv. Mater.* **2009**, *21*, 2151–2155. [CrossRef]
66. Zhang, H.; Son, J.S.; Dolzhnikov, D.S.; Filatov, A.S.; Hazarika, A.; Wang, Y.; Hudson, M.H.; Sun, C.; Chattopadhyay, S.; Talapin, D.V. Soluble lead and bismuth chalcogenidometallates: Versatile solders for thermoelectric materials. *Chem. Mater.* **2017**, *29*, 6396–6404. [CrossRef]
67. Bertino, M.F.; Gadipalli, R.R.; Story, J.G.; Williams, C.G.; Zhang, G.; Sotiriou-Leventis, C.; Tokuhiro, A.T.; Guha, S.; Leventis, N. Laser writing of semiconductor nanoparticles and quantum dots. *Appl. Phys. Lett.* **2004**, *85*, 6007–6009. [CrossRef]
68. Wang, Y.; Fedin, I.; Zhang, H.; Talapin, D.V. Direct optical lithography of functional inorganic nanomaterials. *Science* **2017**, *357*, 385–388. [CrossRef]
69. Wang, Y.; Pan, J.; Wu, H.; Talapin, D.V. Direct wavelength-selective optical and electron-beam lithography of functional inorganic nanomaterials. *ACS Nano* **2019**, *13*, 13917–13931. [CrossRef]
70. Kim, J.; Kwon, S.; Kang, Y.K.; Kim, Y.; Lee, M.; Han, K.; Facchetti, A.; Kim, M.; Park, S.K. A skin-like two-dimensionally pixelized full-color quantum dot photodetector. *Sci. Adv.* **2019**, *5*, eaax8801. [CrossRef]
71. Tang, X.; Chen, M.; Ackerman, M.M.; Melnychuk, C.; Guyot-Sionnest, P. Direct imprinting of quasi-3D nanophotonic structures into colloidal quantum-dot devices. *Adv. Mater.* **2020**, *32*, 1906590. [CrossRef] [PubMed]

72. Ciani, A.J.; Pimpinella, R.E.; Grein, C.H.; Guyot-Sionnest, P. Colloidal quantum dots for low-cost MWIR imaging. In *Infrared Technology and Applications XLII, Proceedings of SPIE Defense + Security, Baltimore, MD, USA, 20 May 2016*; SPIE: Bellingham, WA, USA, 2016; Volume 9819, p. 981919. [CrossRef]
73. Buurma, C.; Pimpinella, R.E.; Ciani, A.J.; Feldman, J.S.; Grein, C.H.; Guyot-Sionnest, P. MWIR imaging with low cost colloidal quantum dot films. In *Optical Sensing, Imaging, and Photon Counting: Nanostructured Devices and Applications 2016, Proceedings of SPIE Nanoscience + Engineering, San Diego, CA, USA,26 September 2016*; SPIE: Bellingham, WA, USA, 2016; Volume 9933, p. 993303. [CrossRef]

Review

CMOS-Compatible Optoelectronic Imagers

Cheng Bi [1,2] and Yanfei Liu [1,*]

[1] Zhongxinrecheng Science and Technology Co., Ltd., Beijing 101102, China
[2] School of Optics and Photonics, Beijing Institute of Technology, Beijing 100081, China
* Correspondence: zxrckj@163.com

Abstract: Silicon-based complementary metal oxide semiconductors have revolutionized the field of imaging, especially infrared imaging. Infrared focal plane array imagers are widely applied to night vision, haze imaging, food selection, semiconductor detection, and atmospheric pollutant detection. Over the past several decades, the CMOS integrated circuits modified by traditional bulk semiconductor materials as sensitivity sensors for optoelectronic imagers have been used for infrared imaging. However, traditional bulk semiconductor material-based infrared imagers are synthesized by complicated molecular beam epitaxy, and they are generally coupled with expensive flip-chip-integrated circuits. Hence, high costs and complicated fabrication processes limit the development and popularization of infrared imagers. Emerging materials, such as inorganic–organic metal halide perovskites, organic polymers, and colloidal quantum dots, have become the current focus point for preparing CMOS-compatible optoelectronic imagers, as they can effectively decrease costs. However, these emerging materials also have some problems in coupling with readout integrated circuits and uniformity, which can influence the quality of imagers. The method regarding coupling processes may become a key point for future research directions. In the current review, recent research progress on emerging materials for infrared imagers is summarized.

Keywords: CMOS; optoelectronic imager; infrared imaging; focal plane array

Citation: Bi, C.; Liu, Y. CMOS-Compatible Optoelectronic Imagers. *Coatings* **2022**, *12*, 1609. https://doi.org/10.3390/coatings12111609

Academic Editor: Saulius Grigalevicius

Received: 16 September 2022
Accepted: 18 October 2022
Published: 22 October 2022

1. Introduction

Complementary metal oxide semiconductors (CMOS) are field-effect transistors that are often used to build integrated circuit chips, such as readout integrated circuits (ROICs), image sensors, and other digital logic circuits. Over the past several decades, an unprecedented change has been noticed in the manufacturing of CMOS. Benefiting from advanced semiconductor manufacturing processes, silicon-based imagers with ultra-high performance (high resolution and wide dynamic response range) can be successfully prepared. Silicon-based imagers have a huge demand in deep learning [1,2], optoelectronic computing [3–5], and neural network computing [6]. The primary reason for the popularity of silicon-based imagers is their compatibility, allowing them to be manufactured on a large scale at a low cost. Due to the success of silicon-based imagers in visible imaging, the main focus of ongoing research is to break through the energy band gap limitation of silicon. Silicon-based CMOS imagers can extend the spectral sensing range for more expensive applications, and broad spectral imagers are important for a wide range of scientific and industrial processes.

Nowadays, the field of infrared silicon-based CMOS imagers is a major area of interest within the field of broad spectral imagers, as infrared contains a wide variety of information in comparison to visible light. The working principle of shortwave infrared imaging (SWIR) is similar to that of visible light imaging. In both processes, images are captured by reflected light, revealing the shadow information of objects. Therefore, information from SWIR imaging can be easily understood. Shortwave infrared can penetrate fog and dust; thus, it can be used in night vision, haze imaging, food selection, and semiconductor detection. In both mid-wave and long-wave infrared imaging, detection is carried out by thermal radiations emitted by targets; therefore, they are extensively applied to autonomous driving [7], gas detection [8], and remote sensing [9]. Studies over the past five decades have provided important information on infrared imagers based on traditional bulk materials. However, traditional bulk materials still have some problems influenced by the imager, such as emerging materials appearing and being applied in the broad spectral imager.

Moreover, the semiconductor industry has also pushed the development of broad-spectrum infrared imagers, which has been noticed in recent years. Focal plane array readout integrated circuits have a better field and a more clever design. Large focal plane array imagers with high-resolution broad-spectrum sensing have attracted significant attention. Benefiting from the development of integrated microcircuits and precise microelectromechanical systems, visible sensing chips can be used in silicon-based broad-spectrum infrared imagers to broaden their sensing spectral range [10]. A wide variety of emerging non-silicon-sensing materials has been successfully integrated with silicon-based readout integrated circuits (ROICs).

In this review, recent research progress on broad-spectrum infrared focal plane area imagers is summarized. The characterization of imager-based focal plane array ROICs is introduced. Then, the history of infrared imagers is simply described, and different infrared optoelectronics imagers are compared with traditional bulk material imagers and emerging material imagers. The current trend of emerging materials and imagers is simply introduced, and the application gap is analyzed. The future research direction of emerging material imagers is also outlooked.

2. Characterization of Focal Plane Area Imagers

Focal plane array (FPA) imagers are special image sensors consisting of an array of optoelectronics-sensitive pixel units on silicon-based chips. Broad-spectrum optoelectronic materials can couple with FPA circuits to prepare broad-spectrum detectors. The pixel structure of a silicon-based FPA imager is displayed in Figure 1, where the center of each pixel dot is a bare electrical connection point with a hole in the passivation layer. These pixel dots can easily transfer signals generated by the optoelectronics material to ROICs. The electronic signal is processed by the operational amplifier circuits and sample-hold devices. Then, the multiplexer selects the pixel data signals and provides an output. Nowadays, FPA ROICs have become a particularly useful choice in fabricating infrared imagers. Tang chose 640 × 512-pixel array ROICs to integrate PbS colloidal quantum dots (CQDs) and build an infrared imager [11]. Anthony et al. designed a mid-wave infrared HgTe CQDs FPA detector [12]. Emmanuel et al. prepared a cost-effective SWIR HgTe CQDs FPA imager [13]. They all chose a similar method by coupling emerging materials and FPA ROICs to prepare their optoelectronic imagers.

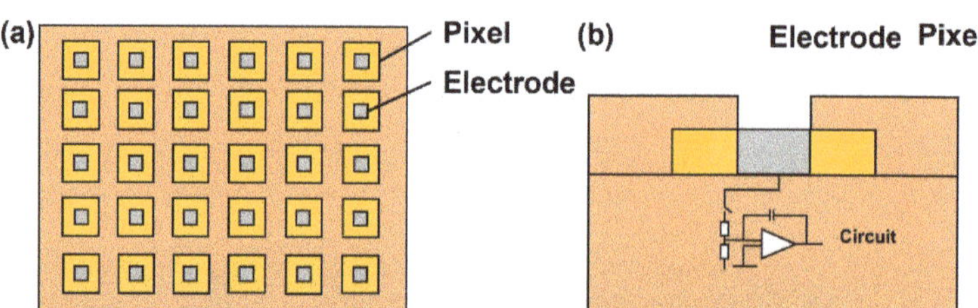

Figure 1. (**a**) Pixel arrangement in a readout circuit (**b**) Single-pixel integrated circuit.

Different from the characterization of single-point optoelectronic devices, FPA imagers use statistical analysis in order to gain insight into the quality of images. The photoresponsivity of imagers is characterized by different parameters, such as photocurrent, dark current, and detectivity. The main parameters of FPA imagers are array and pixel sizes, photoresponse nonuniformity (PRNU), and dead pixel rate. All these parameters are statistically calculated.

The detectivity of FPA imagers is generally affected by spectral responses, the signal-to-noise ratio, response time, and working temperature. Normalized detectivity is calculated as:

$$D^* = \frac{\sqrt{A_{\text{pixel}}}}{V_{\text{rms}} \times \sqrt{2\, t_{\text{integral}}}} \times \frac{V_{\text{ph}}}{P_{\text{in}}}$$

where D^* is the detectivity of an FPA imager, A_{pixel} is the effective size of a pixel, t_{integral} is the integration time for each pixel, V_{ph} is the photoresponsivity of the imager (volt), and V_{rms} is the root mean square of noise.

The external quantum efficiency (EQE) of a photosensitive device is defined as the ratio of incident photons to converted electrons. Generally, a calibrated blackbody is used as a standard light source to irradiate silicon-based FPA imagers; therefore, the EQE of an FPA imager can be calculated as:

$$\text{EQE} = \frac{\text{FWC} \times V_{\text{ph}}/V_{\text{full}}}{\phi_{\text{BB}} \times \Omega \times A_{\text{pixel}} \times t_{\text{integral}}}$$

where FWC is the full well capacity, V_{ph} is the photoresponsivity of the imager (volt), V_{full} is the maximum output voltage, ϕ_{BB} is the photon flux from the blackbody, Ω is the solid angle of the incident photon flux, A_{pixel} is the effective size of a pixel, and t_{integral} is the integration time for each pixel.

Photoresponse nonuniformity (PRNU) is an overlooked fundamental difficulty in manufacturing broad-spectrum silicon-based detectors. The lower the PRUN, the better the image quality, because photoresponse uniformity across the sensing area is the prerequisite for imaging operations. The dead pixel ratio and the hot pixel ratio play vital roles in photoresponse nonuniformity.

In FPA imagers, each pixel on the focal plane is characterized by statistical methods. In order to visualize PRNU, statistical histograms are used to illustrate the response distribution of pixel units in FPA imagers. In statistical histograms, data are binned and summarized by counts or percentages. Each bar in statistical histograms represents a continuous range of data or the frequencies of characteristic data points.

3. Progress of the Infrared Focal Plane Array

With the development of silicon-based integrated circuits, silicon-based readout circuits are widely used in the detection field due to their increased reliability and reduced manufacturing cost. Rapidly advancing science has driven silicon-based integrated circuits into broad-spectrum infrared imaging technology, which has pushed the development of infrared FPA imagers. Readout circuits are designed to collect signals generated by photosensitive materials, ensuring photoresponse uniformity and image quality. In particular, FPA readout circuits have significantly raised the image resolution and detection performance of infrared imaging. Figure 1a exhibits the pixel arrangement in a readout circuit, where hole electrodes are placed in the middle of pixels. A single-pixel integrated circuit is schematically presented in Figure 1b.

Extensive research on broad-spectrum FPA imagers has been executed in recent years. Focal plane detectors in the infrared band can be applied to gas sensing, autonomous driving, and remote sensing. Infrared detectors have high sensitivity to low light and can recognize chemical molecular vibrations and penetrate fog. The efficacy of traditional bulk materials in broad-spectrum infrared FPA imagers is well established. Hence, ongoing research is mainly focusing on infrared-sensitive materials for the preparation of FPA devices [11–14].

Figure 2a presents the invention timeline of different detectors. In 1873, Smith discovered the photoconductive effect of selenium and developed the first photon detector in the world. Kutzscher et al. produced a PbS infrared detector in 1933. PbS infrared detectors were first extensively used in the Second World War, and they could detect shortwave infrared signals of wavelengths up to 3 μm [15]. In the 1950s, infrared detection technology was developed significantly. Cooled infrared detectors, such as lead selenide (PbSe) and indium antimonide (InSb), extended the infrared detect spectrum in the mid-wave infrared range of 3–5 μm. Binary semiconductor materials for infrared detection applications were developed rapidly in this period. In the 1960s, ternary semiconductor compounds emerged. Advanced microelectronics techniques, such as masking, microsoldering, and assembly, were gradually developed, and HgCdTe infrared detectors started to appear.

In 1969, charge-coupled devices (CCDs) were invented. Subsequently, FPA readout circuits were developed for infrared imaging. In the 1970s, the first FPA infrared readout circuit chip was produced [16]. Traditional infrared-sensitive materials have been progressively developed to build staring-vision infrared devices, such as mercury–cadmium–telluride (HgCdTe) devices. The quantum efficiency and array nonuniformity of HgCdTe FPA imagers ranged between 60% and 70% and between 10% and 30%, respectively [17]. In a HgCdTe detector, an infrared detection signal is generated through a thin film of HgCdTe, which is coupled with a flip-chip-integrated readout circuit by molecular beam epitaxy sputtering. A focal plane array readout circuit chip then reads out the analog electrical signal. However, it is very difficult to grow indium columns at each pixel point on the focal plane array and align the HgCdTe material by external stress welding.

Multiquantum well infrared FPA imagers contain a single atomic layer of two different semiconductor materials. GaAs detectors with $D^* > 10^{10}$ Jones can be prepared at working temperatures less than 77 K [18]. In contrast, GaAs multiquantum well infrared FPA imagers have faster response, lower operating power, and better response uniformity. GaAs detectors with large-area readout circuits have a greater potential in the long-wave infrared range than HgCdTe detectors.

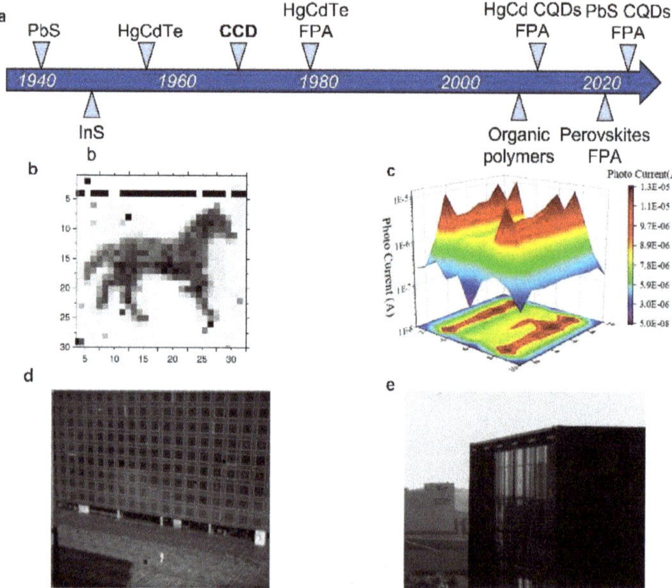

Figure 2. (**a**) Invention timeline of different infrared detectors; (**b**) image taken by a hybrid CMOS detector equipped with a NIR-light-emitting diode at 850 nm [19]. Copyright 2012, Nature Communications; (**c**) Number "11" captured by an infrared perovskite detector [20]. Copyright 2019, Advanced Materials Technologies; (**d**) SWIR image of a building (taken with a HgTe CQD detector) [13]. Copyright 2022, Nanoscale.; (**e**) photo of outdoor buildings at nightfall (captured by a PbS CQD imager) [11]. Copyright 2022, Nature Electronics.

The cut-off wavelength and response nonuniformity of InSb focal plane array detectors at 77 K are 5 μm and 10%, respectively. Generally, InSb films can be easily prepared in large-area readout circuits by a mature process. Although InSb FPA imagers and HgCdTe FPA detectors have a similar scale and pixel uniformity, HgCdTe detectors have a faster response and higher quantum efficiency during MWIR detection at low temperatures. However, InSb has good potential for MWIR detection because of its simple crystal growth process and fabrication reliability.

Nowadays, traditional bulk optoelectronic materials have been widely applied to broad-spectrum imagers. However, traditional infrared optoelectronics imagers still have some problems, which are that it is difficult to solve and influence the cost and yield of imagers, as traditional bulk semiconductor materials require lattice-matched substrates and complex flip-bonding processes to get attached to silicon-integrated readout circuits (ROICs), such as HgCdTe, InGaAs, and InSb. The detector manufacturing cost increases dramatically with larger arrays and smaller pixel sizes [21]. Complex processes have impeded the development of high-resolution FPA detectors. In order to further expand the array scale and sensing range of Si-CMOS imagers, researchers are trying to replace the flip-bonding process with the copper–copper bonding technique and germanium–silicon technique [22,23]; however, this approach cannot yet solve the bonding problem between traditional block infrared-sensitive materials and silicon-based CMOS readout circuits.

The coupling of conventional block materials to readout circuits is mechanically unfeasible. The difficulty of integrating nonsilicon materials into silicon-integrated circuits has greatly hindered the development of Si-CMOS imagers beyond visible light. Emerging materials, such as inorganic–organic metal halide perovskites [24–28], organic polymers [19,29,30], and colloidal quantum dots (CQDs) [12,14,31–37], can avoid complex flip-chip bonding processes to reduce the detector production cost and increase the portability of FPA readout circuit detectors. It is hoped that these emerging optoelectronic materials will contribute to solving the problem brought by traditional bulk materials and promote the application of infrared imagers.

As shown in Figure 3, Daniela Baierl developed a hybrid CMOS detector based on a bulk heterojunction structure of PCBM–P3HT (poly(3-hexylthiophene)–[6, 6]-phenyl C61 butyric acid methylester) and squaraine to enable detection in the near-infrared range [19]. The detector has a fill factor of 100% and an EQE of up to 50%. Figure 2b displays a horse image taken by a hybrid CMOS imager under NIR light-emitting diode illumination at 850 nm. The hybrid CMOS detector is prepared by a low-cost spray-coating process to deposit the optoelectronic organic polymer film on the CMOS ROIC. It offers an effective method of coupling with integrated circuits. However, this organic polymer imager has some limitations in response range and detectivity. There is, therefore, a definite need for finding a great organic polymer material for application in the optoelectronics field.

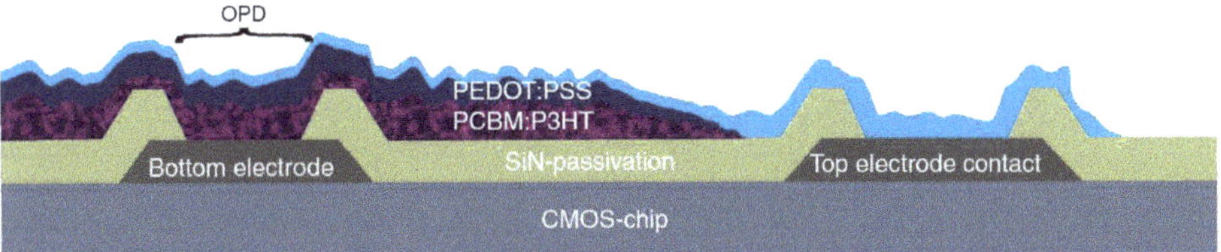

Figure 3. Structure of hybrid CMOS detector [19]. Copyright 2012, Nature Communications.

Inorganic–organic metal halide perovskites play an important role in infrared imaging. Wang fabricated an infrared imager by depositing a low-bandgap Sn-Pb-based perovskite photodiode on an indium gallium zinc oxide thin-film transistor (TFT) [20]. A layer of hydrophobic perfluoro (1-butenyl vinyl ether) polymer (CYTOP) was printed on the TFT (Figure 4). A $(FASnI_3)_{0.6}(MAPbI_3)_{0.4}$ perovskite photodiode was deposited on the pixel electrode by spin coating, and the solution was self-assembled due to the hydrophobicity of the CYTOP layer. The inorganic–organic metal halide perovskite imager had 12 × 12 matrix arrays with an ITO/SnO_2/$(FASnI_3)_{0.6}(MAPbI_3)_{0.4}$/P3HT/Au multilayered photodiode structure. The low-bandgap perovskite imager had a wide response range of 300–1000 nm, and its EQE and detectivity at 850 nm were 45% and 10^{11} Jones, respectively. However, the response of the perovskite imager was nonuniform, and only half of the pixel electrode worked as a photodiode in 12 × 12 matrix arrays, and this phenomenon could be attributed to the existence of the uneven perovskite film. The defected perovskite film provided a current leakage path. The number "11" shown in Figure 2c is captured by a perovskite photodiode. Inorganic–organic metal halide perovskite coupling with TFT circuits is an interesting topic. This study contributes to our understanding of the application of halide perovskites on the imager. Inorganic–organic metal halide perovskite imagers should improve the uniformity and crystallinity of the perovskite layer to enhance the performance of the optoelectronic response, which increases the difficulty of the process, and the device interface between the perovskite film and circuit also influences the dark current and the uniformity of the pixel response. Greater efforts are needed to improve the interface morphology of TFT circuits and the uniformity of perovskite film.

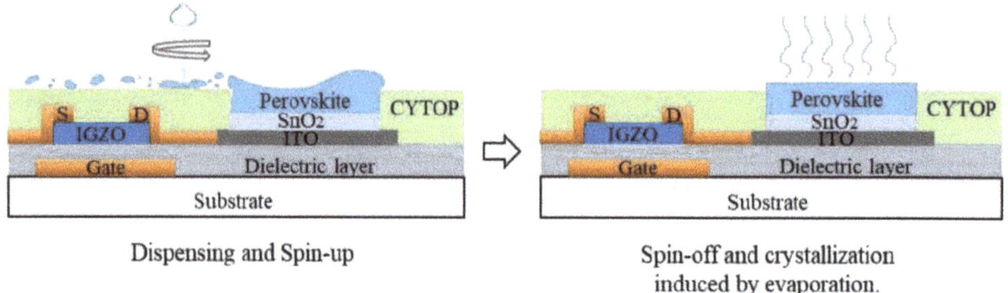

Figure 4. Perovskite layer coupled with TFT arrays by spin-on-pattern deposition [20]. Copyright 2019, Advanced Materials Technologies.

The most surprising materials in the infrared emerging materials field are CQDs. A variety of CQDs are now available to extend the range of infrared detection [38]. The development of low-cost FPA detectors based on CQDs is achieved by coupling CQDs with readout circuits [39]. Tang built an infrared imager by integrating PbS CQDs on the surface of a Si-CMOS readout circuit [11]. The photodetector was integrated into a CMOS ROIC with a 640 × 512-pixel array, a pixel size of 15 μm, an EQE of 60%, and a spatial resolution of 40 line pairs per millimeter at a modulation transfer function of 50%. The detector was composed of a multilayered indium tin oxide (ITO)/NiO_x/PbS CQD/C_{60}/ZnO/ITO (Figure 5a,b). Generally, photodiodes of PbS CQDs FPA imagers are composed of NiO_x/PbS CQD/C_{60}/ZnO. Charge transport layers (NiO_x and ZnO) are prepared by sputtering, and a protective layer of fullerene (C_{60}) is deposited on the PbS CQD/ZnO interface. NiO_x and ZnO act as hole transport and electron transport layers, respectively, to provide suitable energy band positions and high stability. In order to improve reproducibility, hole transport layers and electron transport layers are prepared by mild sputtering. Fullerene protective layers, with their strong C-C bonding and mechanical properties, can prevent damage during sputtering. In the PbS imager shown in Figure 5a, a transparent conductive oxide (ITO) is used as the electrode. The device could collect more charges in the top-illuminated mode than in the back-illuminated mode. The depletion region of the photodiode close to the illumination side has prompted the efficient drift of photogenerated carriers by the built-in electric field. Figure 2e exhibits an image of outdoor buildings at nightfall captured by a PbS quantum dot device. The most obvious finding to emerge from this study is that the CQD imager can take clear images, and CQD material has more mature applications coupling with ROICs. However, one source of weakness in this study, which could have affected the measurements of infrared, was the PbS CQD imager exhibits a spectral range of 400–1300 nm, which cannot fully cover the range of SWIR.

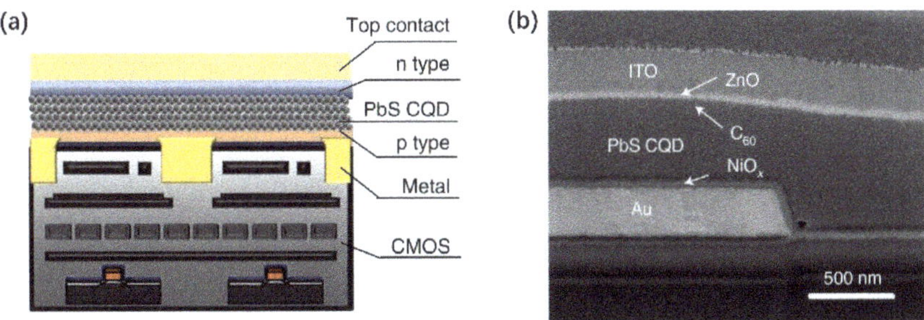

Figure 5. (**a**) Schematic diagram and (**b**) SEM image of the cross-section of a PbS imager [11]. Copyright 2022, Nature Electronics.

Further, HgTe CQDs are band gap tunable quantum dot materials. The special properties of HgTe CQDs reduce complexities in manufacturing and extend the spectral response range. Controlling the quantum dot size, HgTe CQDs can be used to detect a wide range of infrared wavelengths, such as shortwave infrared, mid-wave infrared, and long-wave infrared [31]. HgTe CQDs FPA infrared detectors can operate at higher temperatures than traditional detectors based on HgCdTe, PbTe, and In(Ga)AsSb [40]. Guyot and Sionnest invented a very sophisticated technique to synthesize HgTe CQDs using cheap precursors in less than two hours [12]. Colloidal suspensions of HgTe CQDs can be synthesized by the bench-top method and directly coupled with readout circuits by spraying, dripping, and spin coating, avoiding complex molecular epitaxial growth and flip-bonding processes. Anthony designed a mid-wave infrared HgTe CQDs FPA detector. The QE, D^*, and NEDT of the MWIR HgTe CQD FPA imager at 95 K were 0.30%, 1.46×10^9 Jones, and 2.319, respectively. Defects in CQD films can greatly reduce the performance of detectors. However, ROIC modification and CQD deposition processes still need to be improved to enhance the performance of detectors.

Emmanuel prepared a cost-effective HgTe CQDs focal plane array by a single-step fabrication route [13]. Figure 2d displays an image of a building captured by an SWIR HgTe CQDs device. The HgTe CQDs film deposited on a 15 μm pixel FPA readout circuit has an EQE of 4%–5% and a cut-off wavelength of 1.8 μm in the Peltier cooling condition. In Figure 6a, an ITO-covered glass and a silicon wafer are placed in front of different reagent vials. However, in Figure 6b, an opaque image of the ITO glass and the reagents and a transparent image of the silicon wafer are exhibited. This is because absorption bands of silicon range between 0.4 and 1.1 μm, and the transmissivity of the ITO glass ranges between 0.4 and 0.9 μm. A single imager generally costs EUR 70. This single-step fabrication route can avoid the expensive molecular beam epitaxy method. Table 1 lists the parameters of infrared detectors based on emerging materials. Compared with the other emerging materials, HgTe CQDs as band gap tunable material can extend spectral the range of response spectral by controlling the condition of the experiment, and HgTe CQDs can be uniformly modified on FPA circuits, which improves the quality of images and decrease the cost of the process. However, there are some problems that should be solved, such as the agglomeration of CQDs influencing the response of pixels. Notwithstanding these limitations, the study suggests that HgTe CQDs will make an important contribution to the field of broad-spectrum optoelectronic imagers.

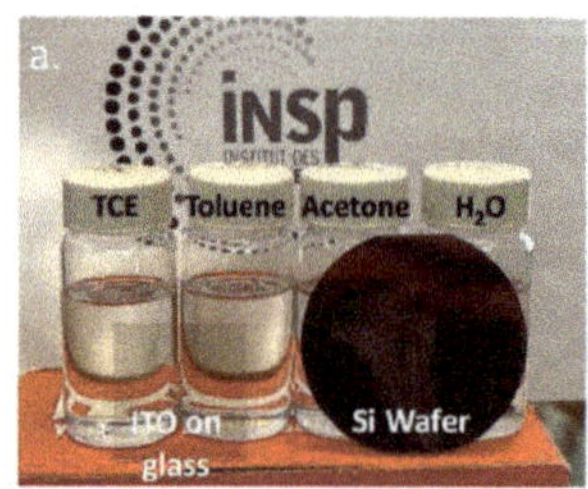

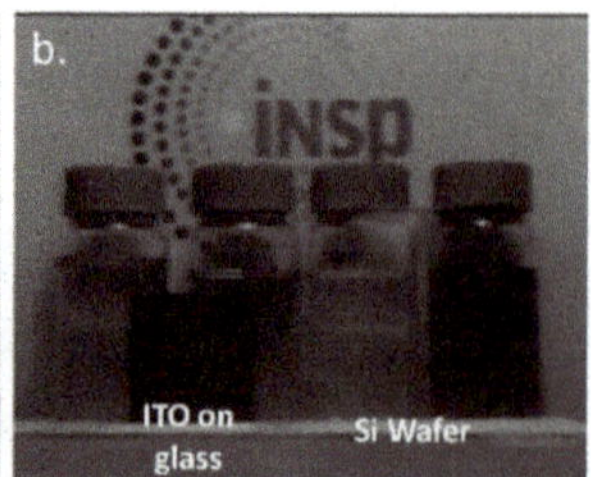

Figure 6. (**a**) Complicated arrangement of different reagent vials, an ITO glass, and a silicon wafer under visible light and (**b**) SWIR image of the complicated arrangement. [13]. Copyright 2022, Nanoscale.

Table 1. Parameters of infrared detectors based on emerging materials.

No.	Detector Type	Pixel Sizes (μm)	Wavelength (μm)	EQE (%)	PRNU (%)	Detectivity (Jones)	Resolution (px)
1	PCBM: P3HT organic polymers	15,000	0.6–0.9	50	N/A	N/A	30 × 30
2	$(FASnI_3)_{0.6}(MAPbI_3)_{0.4}$ perovskite photodiode	N/A	0.3–1.0	45	N/A	1×10^{11}	12 × 12
3	PbS CQDs photodiodes	15	0.7–1.4	63	3.38	2.1×10^{12}@RT	640 × 512
4	HgTe photoconductor (MWIR)	30	0.8–3.6	N/A	N/A	1.25×10^{11}	320 × 256
5	HgTe photoconductor (SWIR)	30	0.8–1.8	4–5	N/A	N/A	640 × 512

4. Outlook

Si-CMOS technology can successfully fabricate high-performance and low-cost detectors with high reliability. Si-CMOS readout circuits facilitate the development of infrared FPA imagers. Staring-vision readout circuits are extensively used in the infrared field. Current research mainly focuses on the development of large-array infrared detectors with high resolution and low noise. Infrared CMOS-compatible optoelectronic imagers have a wide application in the electronics industry, ranging from vibrating chemical molecule detection to haze imaging. Broad-spectrum infrared imagers are used in semiconductor detection, medical drug screening, and agricultural product observation. However, infrared imagers have some major issues that need to be resolved before their mass rollout.

Traditional infrared FPA imagers are maturing; however, complex fabrication processes and high manufacturing cost limit their development. Generally, indium columns are grown in the center of pixel electrodes. Block materials are then coupled to focal plane array readout circuits under external pressure. This process seriously affects the product quality of infrared imagers. When readout circuits are scaled up, the difficulty of manufacturing significantly increases. New encapsulation methods might increase the portability of block materials.

New optoelectronic materials, such as perovskites, organic films, and colloidal quantum dots, are impetuses to the development of infrared FPA imagers. Due to their flowability, these new optoelectronic materials can easily flow to each pixel electrode on the surface of FPA readout circuits, and this simple coupling method avoids complex processes. In comparison to bulk materials, these new materials can reduce manufacturing costs and extend the application field of detectors. Therefore, significant research is being carried out on these new optoelectronic materials to build infrared FPA imagers with high stability and strong sensitivity. However, coupling processes between these new materials and readout circuits need further investigation.

Author Contributions: Writing—original draft preparation, C.B.; writing—review and editing, Y.L. All authors have read and agreed to the published version of the manuscript.

Funding: This research received no external funding.

Institutional Review Board Statement: Not applicable.

Informed Consent Statement: Not applicable.

Data Availability Statement: Not applicable.

Conflicts of Interest: The authors declare no conflict of interest.

References

1. Nakadai, M.; Asano, T.; Noda, S. Electrically controlled on-demand photon transfer between high-Q photonic crystal nanocavities on a silicon chip. *Nat. Photonics* **2021**, *16*, 113–118. [CrossRef]
2. Arakawa, Y.; Nakamura, T.; Urino, Y.; Fujita, T. Silicon photonics for next generation system integration platform. *IEEE Commun. Mag.* **2013**, *51*, 72–77. [CrossRef]
3. Liu, W.; Li, M.; Guzzon, R.S.; Norberg, E.J.; Parker, J.S.; Lu, M.; Coldren, L.A.; Yao, J. A fully reconfigurable photonic integrated signal processor. *Nat. Photonics* **2016**, *10*, 190–195. [CrossRef]
4. Zhou, T.; Lin, X.; Wu, J.; Chen, Y.; Xie, H.; Li, Y.; Fan, J.; Wu, H.; Fang, L.; Dai, Q. Large-scale neuromorphic optoelectronic computing with a reconfigurable diffractive processing unit. *Nat. Photonics* **2021**, *15*, 367–373. [CrossRef]
5. Piggott, A.Y.; Lu, J.; Lagoudakis, K.G.; Petykiewicz, J.; Babinec, T.M.; Vučković, J. Inverse design and demonstration of a compact and broadband on-chip wavelength demultiplexer. *Nat. Photonics* **2015**, *9*, 374–377. [CrossRef]
6. Shastri, B.J.; Tait, A.N.; Ferreira de Lima, T.; Pernice, W.H.; Bhaskaran, H.; Wright, C.D.; Prucnal, P.R. Photonics for artificial intelligence and neuromorphic computing. *Nat. Photonics* **2021**, *15*, 102–114. [CrossRef]
7. Jiang, Y.; Karpf, S.; Jalali, B. Time-stretch LiDAR as a spectrally scanned time-of-flight ranging camera. *Nat. Photonics* **2020**, *14*, 14–18. [CrossRef]
8. Muraviev, A.; Smolski, V.; Loparo, Z.; Vodopyanov, K. Massively parallel sensing of trace molecules and their isotopologues with broadband subharmonic mid-infrared frequency combs. *Nat. Photonics* **2018**, *12*, 209–214. [CrossRef]
9. Liu, J.; Dai, J.; Chin, S.L.; Zhang, X.-C. Broadband terahertz wave remote sensing using coherent manipulation of fluorescence from asymmetrically ionized gases. *Nat. Photonics* **2010**, *4*, 627–631. [CrossRef]
10. Aziz, N.Y.; Kincaid, G.T.; Parrish, W.J.; Woolaway II, J.T.; Heath, J.L. Standardized high-performance 320 by 256 readout integrated circuit for infrared applications. In Proceedings of the Infrared Readout Electronics IV, Orlando, FL, USA, 13 April 1998; pp. 80–90.
11. Liu, J.; Liu, P.; Chen, D.; Shi, T.; Qu, X.; Chen, L.; Wu, T.; Ke, J.; Xiong, K.; Li, M.; et al. A near-infrared colloidal quantum dot imager with monolithically integrated readout circuitry. *Nat. Electron.* **2022**, *5*, 443–451. [CrossRef]
12. Ciani, A.J.; Pimpinella, R.E.; Grein, C.H.; Guyot-Sionnest, P. Colloidal quantum dots for low-cost MWIR imaging. In Proceedings of the SPIE Defense + Security, Baltimore, MD, USA, 20 May 2016.
13. Gréboval, C.; Darson, D.; Parahyba, V.; Alchaar, R.; Abadie, C.; Noguier, V.; Ferré, S.; Izquierdo, E.; Khalili, A.; Prado, Y.; et al. Photoconductive focal plane array based on HgTe quantum dots for fast and cost-effective short-wave infrared imaging. *Nanoscale* **2022**, *14*, 9359–9368. [CrossRef] [PubMed]
14. Zhang, S.; Hu, Y.; Hao, Q. Advances of sensitive infrared detectors with HgTe colloidal quantum dots. *Coatings* **2020**, *10*, 760. [CrossRef]
15. Rogalski, A. History of infrared detectors. *Opto-Electron. Rev.* **2012**, *20*, 279–308. [CrossRef]
16. Boyle, W.S.; Smith, G.E. Charge coupled semiconductor devices. *Bell Syst. Technol. J.* **1970**, *49*, 587–593. [CrossRef]
17. Pultz, G.; Norton, P.W.; Krueger, E.E.; Reine, M. Growth and characterization of P-on-n HgCdTe liquid-phase epitaxy heterojunction material for 11–18 μm applications. *AIP Conf. Proc.* **1991**, *235*, 1724–1730.
18. Gunapala, S.; Bandara, K. Recent developments in quantum-well infrared photodetectors. *Thin Film.* **1995**, *21*, 262–275. [CrossRef]
19. Baierl, D.; Pancheri, L.; Schmidt, M.; Stoppa, D.; Dalla Betta, G.-F.; Scarpa, G.; Lugli, P. A hybrid CMOS-imager with a solution-processable polymer as photoactive layer. *Nat. Commun.* **2012**, *3*, 1175. [CrossRef] [PubMed]
20. Wang, Y.; Chen, C.; Zou, T.; Yan, L.; Liu, C.; Du, X.; Zhang, S.; Zhou, H. Spin-On-Patterning of Sn–Pb Perovskite Photodiodes on IGZO Transistor Arrays for Fast Active-Matrix Near-Infrared Imaging. *Adv. Mater. Technol.* **2019**, *5*, 1900752. [CrossRef]
21. Rogalski, A.; Antoszewski, J.; Faraone, L. Third-generation infrared photodetector arrays. *J. Appl. Phys.* **2009**, *105*, 091101. [CrossRef]
22. Sood, A.K.; Richwine, R.A.; Puri, Y.R.; DiLello, N.A.; Hoyt, J.L.; Akinwande, T.; Dhar, N.K.; Horn, S.B.; Balcerak, R.S.; Bramhall, T.G. Development of low dark current SiGe-detector arrays for visible-NIR imaging sensor. In Proceedings of the Defense, Security, and Sensing, Orlando, FL, USA, 6 May 2009.
23. Manda, S.; Zaizen, Y.; Hirano, T.; Iwamoto, H.; Matsumoto, R.; Saito, S.; Maruyama, S.; Minari, H.; Takachi, T.; Fujii, N.; et al. High-definition Visible-SWIR InGaAs Image Sensor using Cu-Cu Bonding of III-V to Silicon Wafer. In Proceedings of the 2019 IEEE International Electron Devices Meeting (IEDM), San Francisco, CA, USA, 7–11 December 2019; pp. 16.17.11–16.17.14. [CrossRef]

24. Deumel, S.; van Breemen, A.J.J.M.; Gelinck, G.G.; Peeters, B.; Maas, J.; Verbeek, R.; Shanmugam, S.K.; Akkerman, H.; Meulenkamp, E.A.; Huerdler, J.E.; et al. High-sensitivity high-resolution X-ray imaging with soft-sintered metal halide perovskites. *Nat. Electron.* **2021**, *4*, 681–688. [CrossRef]
25. Kim, Y.C.; Kim, K.H.; Son, D.-Y.; Jeong, D.-N.; Seo, J.-Y.; Choi, Y.S.; Han, I.T.; Lee, S.Y.; Park, N.G. Printable organometallic perovskite enables large-area, low-dose X-ray imaging. *Nature* **2017**, *550*, 87–91. [CrossRef]
26. Wei, W.; Zhang, Y.; Xu, Q.; Wei, H.; Fang, Y.; Wang, Q.; Deng, Y.; Li, T.; Gruverman, A.; Cao, L.R.; et al. Monolithic integration of hybrid perovskite single crystals with heterogenous substrate for highly sensitive X-ray imaging. *Nat. Photonics* **2017**, *11*, 315–321. [CrossRef]
27. Wang, D.; Li, G. Advances in photoelectric detection units for imaging based on perovskite materials. *Laser Photonics Rev.* **2022**, *16*, 2100713. [CrossRef]
28. Jana, A.; Bathula, C.; Park, Y.; Kadam, A.N.; Sree, V.G.; Ansar, S.; Kim, H.-S.; Im, H. Facile synthesis and optical study of organic-inorganic lead bromide perovskite-clay (kaolinite, montmorillonite, and halloysite) composites. *Surf. Interfaces* **2022**, *29*, 101785. [CrossRef]
29. Bathula, C.; Opoku, H.; Gopalan Sree, V.; Kadam, A.N.; Meena, A.; Reddy Palem, R.; Misra, M.; Naushad, M.; Im, H.; Kim, H.-S. Optoelectronic and DFT investigation of thienylenevinylene based materials for thin film transistors. *J. Mol. Liq.* **2022**, *360*, 119462. [CrossRef]
30. Bathula, C.; Rabani, I.; Opoku, H.; Youi, H.-K.; Gopal Sree, V.; Mane, S.D.; Seo, Y.-S.; Kim, H.S. Efficient synthesis of acetylene-bridged carbazole-based dimer for electrochemical energy storage: Experimental and DFT studies. *J. Electroanal. Chem.* **2021**, *889*, 115225. [CrossRef]
31. Buurma, C.; Pimpinella, R.E.; Ciani, A.J.; Feldman, J.S.; Grein, C.H.; Guyot-Sionnest, P. MWIR imaging with low cost colloidal quantum dot films. In Proceedings of the NanoScience + Engineering, San Diego, CA, USA, 26 September 2016.
32. Kim, J.H.; Pejović, V.; Georgitzikis, E.; Li, Y.; Kim, J.; Malinowski, P.E.; Lieberman, I.; Cheyns, D.; Heremans, P.; Lee, J. Detailed characterization of short-wave infrared colloidal quantum dot image sensors. *IEEE Trans. Electron Devices* **2022**, *69*, 2900–2906. [CrossRef]
33. Georgitzikis, E.; Malinowski, P.E.; Li, Y.; Maes, J.; Hagelsieb, L.M.; Guerrieri, S.; Hens, Z.; Heremans, P.; Cheyns, D. Integration of PbS quantum dot photodiodes on silicon for NIR imaging. *IEEE Sens. J.* **2020**, *20*, 6841–6848. [CrossRef]
34. Pejović, V.; Georgitzikis, E.; Lieberman, I.; Malinowski, P.E.; Heremans, P.; Cheyns, D. Photodetectors based on lead sulfide quantum dot and organic absorbers for multispectral sensing in the visible to short-wave infrared range. *Adv. Funct. Mater.* **2022**, *32*, 2201412. [CrossRef]
35. Pejović, V.; Lee, J.; Georgitzikis, E.; Li, Y.; Kim, J.H.; Lieberman, I.; Malinowski, P.E.; Heremans, P.; Cheyns, D. Thin-film photodetector optimization for high-performance short-wavelength infrared imaging. *IEEE Electron Device Lett.* **2021**, *42*, 1196–1199. [CrossRef]
36. Steckel, J.S.; Josse, E.; Pattantyus-Abraham, A.G.; Bidaud, M.; Mortini, B.P.; Bilgen, H.; Arnaud, O.; Allegret-Maret, S.; Saguin, F.; Mazet, L.; et al. 1.62 μm Global Shutter Quantum Dot Image Sensor Optimized for Near and Shortwave Infrared. In Proceedings of the 2021 IEEE International Electron Devices Meeting (IEDM), San Francisco, CA, USA, 11–16 December 2021; pp. 23.24.21–23.24.24. [CrossRef]
37. Zhao, P.; Mu, G.; Chen, M.; Tang, X. Simulation of resonant cavity-coupled colloidal quantum-dot detectors with polarization sensitivity. *Coatings* **2022**, *12*, 499. [CrossRef]
38. Pejović, V.; Georgitzikis, E.; Lee, J.; Lieberman, I.; Cheyns, D.; Heremans, P.; Malinowski, P.E. Infrared colloidal quantum dot image sensors. *IEEE Trans. Electron Devices* **2022**, *69*, 2840–2850. [CrossRef]
39. Allen, M.; Bessonov, A.D.; Ryhänen, T. 66-4: Invited paper: Graphene enhanced QD image sensor technology. *SID Symp. Dig. Technol. Pap.* **2021**, *52*, 987–990. [CrossRef]
40. Dong, Y.; Chen, M.; Yiu, W.K.; Zhu, Q.; Zhou, G.; Kershaw, S.V.; Ke, N.; Wong, C.P.; Rogach, A.L.; Zhao, N. Solution processed hybrid polymer: HgTe quantum dot phototransistor with high sensitivity and fast infrared response up to 2400 nm at room temperature. *Adv. Sci.* **2020**, *7*, 2000068. [CrossRef] [PubMed]

Review

Room-Temperature Infrared Photodetectors with Zero-Dimensional and New Two-Dimensional Materials

Taipeng Li [1], Xin Tang [1,2,3] and Menglu Chen [1,2,3,*]

[1] School of Optics and Photonics, Beijing Institute of Technology, Beijing 100081, China; 3120210626@bit.edu.cn (T.L.); xintang@bit.edu.cn (X.T.)
[2] Beijing Key Laboratory for Precision Optoelectronic Measurement Instrument and Technology, Beijing 100081, China
[3] Yangtze Delta Region Academy of Beijing Institute of Technology, Jiaxing 314019, China
* Correspondence: menglu@bit.edu.cn

Abstract: Infrared photodetectors have received much attention for several decades due to their broad applications in the military, science, and daily life. However, for achieving an ideal signal-to-noise ratio and a very fast response, cooling is necessary in those devices, which makes them bulky and costly. Thus, room-temperature infrared photodetectors have emerged as a hot research direction. Novel low-dimensional materials with their easy fabrication and excellent photoelectronic properties provide a possible solution for room-temperature infrared photodetectors. This review aims to summarize the preparation methods and characterization of several low-dimensional materials (PbS, PbSe and HgTe, new two-dimensional materials) with great concern and the room-temperature infrared photodetectors based on them.

Keywords: infrared photodetectors; room temperature; low-dimensional

Citation: Li, T.; Tang, X.; Chen, M. Room-Temperature Infrared Photodetectors with Zero-Dimensional and New Two-Dimensional Materials. *Coatings* **2022**, *12*, 609. https://doi.org/10.3390/coatings12050609

Academic Editor: Alexandru Enesca

Received: 20 March 2022
Accepted: 27 April 2022
Published: 29 April 2022

1. Introduction

In nature, every object above absolute zero can emit infrared photons with unique fingerprints of different temperature information, though they cannot be captured by human eyes. Infrared detectors, however, can convert infrared radiation into a measurable electronic signal, which is significant in infrared technology [1]. The technical progress of infrared technology is mainly related to the development of narrow-band semiconductor infrared photodetectors, which have both an ideal signal-to-noise ratio and a very fast response. However, to achieve this goal, the photodetectors need to be cooled at low temperatures to avoid thermal-activated carriers. As a result, the system is bulky, costly and inconvenient to use. As a result, developing room-temperature infrared photodetectors has become an important research direction [1].

Driven by Moore's Law, the feature size of highly integrated electronics has already been reduced to a few nanometers [2]. Consequently, in infrared technology, one major focus is how to reduce pixel scale and increase format array for a highly integrated photoelectric detection system with high performance. Traditional infrared photodetectors based on narrow-band semiconductors, such as indium–gallium arsenide (InGaAs) and mercury–cadmium telluride (HgCdTe), need the flip-up bonding technique. This causes great difficulties in coupling with silicon electronics, limiting the development of the focal plane array (FPA). Low-dimensional nanostructured materials are appealing in the field of photoelectric detection because they can be integrated with traditional silicon electronics and even with flexible large-area substrates by liquid phase processing, such as spin-coating, spraying, or stratified deposition [3]. In addition, they have great potential in subwavelength pixel, large array and multicolor devices [4]. However, there are still many challenges in material growth, device fabrication and coupling with circuits. The development of low-dimensional material infrared photodetectors started 40 years ago. In recent times, most of the reported infrared detectors, such as those based on mercury telluride (HgTe) colloidal quantum dots (CQDs), are still single-pixel devices [5]. Figure 1 shows the history of the development of infrared detectors and room-temperature infrared detectors with low-dimensional materials.

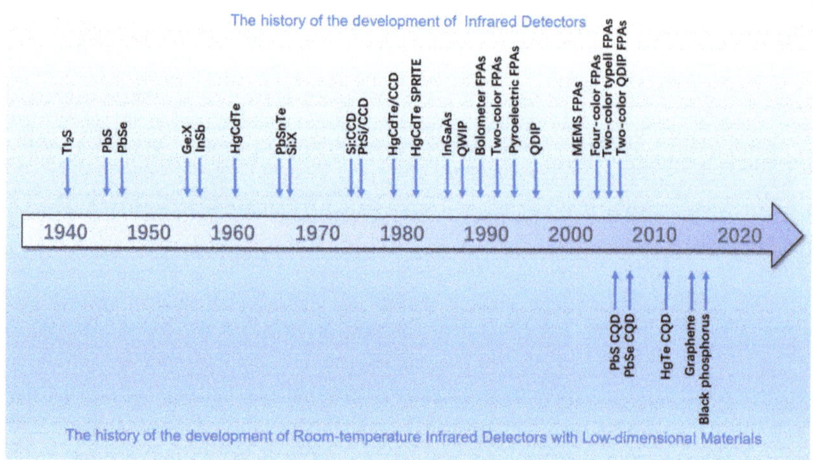

Figure 1. "The history of the development of infrared detectors" and "The history of the development of room-temperature infrared detectors with low-dimensional materials".

Low-dimensional materials can be classified into zero-dimensional (0D), one-dimensional (1D) and two-dimensional (2D) materials.

CQDs, which are typically 2–20 nm in diameter, are important representatives of 0D materials [6]. Benefiting from quantum confinement, the band gap of CQDs is strongly dependent on nanocrystal (NC) size, which makes it possible to design a band gap independently. This means that, by controlling the size of CQDs, rather than their chemical composition, a wide spectrum can be covered [7]. For example, the spectra of HgTe CQD can cover short-wave infrared (SWIR), mid-wave infrared (MWIR), long-wave infrared (LWIR) and even Tera Hertz (THz) [8]. Due to the rapid development of material systems, it is now even possible to achieve this wide spectral range using non-toxic and heavy metal-free quantum dots [9]. In addition, the preparation process of CQDs has significant advantages. CQDs are typically made from II-IV, III-V and IV-VI semiconductors through inexpensive and scalable wet chemical synthesis [6]. This enables cheap large-scale manufacturing of optoelectronic equipment at room temperature by several techniques, such as roll-to-roll processing [7]. The flexibility of CQDs is also reflected in the fact that the modern solution of ligand exchange protocols can change the properties of CQDs prior to the deposition process. For example, long-chain organic ligands on the surface of CQDs can be replaced by short-chain organic compounds and even by inorganic ions such as bromides or iodides for better carrier transport [10]. In fact, CQDs have already achieved great success in high-performance infrared detection and in multi-functions such as those in SWIR/MWIR dual band photodetectors [11] and polarization infrared detectors [12] due to unique structure-related photoelectric properties.

In recent years, 1D and 2D nanomaterials have become the focus of nanotechnology research. At present, 1D nanomaterials can be synthesized from a single crystal form, whose chemical composition, shape, doping state, length, diameter and other key parameters are controllable [13]. Now, near-infrared (NIR) photodetectors based on MBE-grown InAs NWs show high-performance optoelectronic properties. Single InAs NW photodetectors have an impressive I_{on}/I_{off} ratio of 10^5 with a maximum field-effect mobility of $\sim$2000 cm^2 /V·s [14]. For 2D materials, graphene is one main research direction in infrared detectors.

A waveguide-integrated graphene photodetector that simultaneously exhibits high responsivity, high speed and broad spectral band width can achieve a photoresponsivity exceeding 0.1 AW^{-1} together with a nearly uniform response between 1450 and 1590 nm [15]. Another 2D material of interest is black phosphorus. Black phosphorus MWIR detectors have been demonstrated at 3.39 μm with high internal gain, resulting in an external responsivity of 82 A/W [16]. In addition, there is a lot of research on the physicochemical properties of 2D electron gas [17–19].

Excellent light response and signal-to-noise ratios are indispensable for obtaining an excellent room-temperature detector. For PC devices, 1/f noise and shot noise are inevitable because of bias voltage. For PV devices, 1/f noise and Johnson noise are the main interference signals when there is no bias voltage, whereas shot noise has a greater effect than Johnson noise when bias voltage is applied.

In this review, we excessively introduce room-temperature infrared detectors based on lead sulfide (PbS), lead selenide (PbSe) and HgTe CQDs, which belong to 0D materials, as well as graphene and black phosphorus, which belong to 2D materials. There have been good review articles on room-temperature detectors and materials before. The review of PbSe thin-film infrared detectors reported by Gupta et al. in 2021 [20] introduces the physical and chemical properties of this material in detail and summarizes the latest progress in the research of PbSe photoconductivity. The review reported by Wang et al. in 2019, which introduces mature technology of large-scale commercialization, newly developed technology based on quantum cascade photodetectors, and brand-new concept devices based on 2D materials, provides a comprehensive view of the progress and challenges of room-temperature infrared detectors [21]. Reviews of room-temperature photoelectric infrared detectors focusing only on a certain material have more detailed descriptions of the properties and applications of this material, such as a review reported in 2019 which focuses on a hybrid structure based on 2D materials [22]. A review reported in 2018, which summarizes emerging technologies for high performance IR detectors, takes the material type as the framework, which is similar to our review [23]. This review aims to cover more materials for room-temperature IR photodetectors.

2. PbS

The development history of PbS and PbSe is introduced here because the developments of various lead salt materials are closely related to each other. As a result, the development course of PbSe is not repeated in Section 3.

The earliest information about lead-based semiconductor materials comes from a patent published in 1904 by Bose, who found and utilized the photovoltaic effect of a crystal of galena. Subsequently, Case carried out his research on thin films of thallous sulfide (Tl_2S) in 1917 [24] and 1920 [25]. Due to the military needs of infrared information in World War II, Germany developed lead salt (PbS, PbSe and lead telluride (PbTe)) materials vigorously in the 1930s. During that period, different methods for preparing lead salt thin films developed rapidly. Gudden and Kutzscher prepared lead salt films by evaporation and chemical deposition, respectively. Shortly after German scientists firstly studied it, the United States scientists also conducted research on it. Cashman of Northwestern University began work on Tl_2S in 1941 and later turned his full attention to the preparation of thin films of PbS, PbSe and PbTe by vacuum evaporation. Among the three typical lead salts used in infrared detectors, PbS and PbSe have been developed and produced to some extent, but PbTe has not been adapted for production and has been gradually phased out [26]. The cut-off wavelength of PbS and PbSe QD materials is short wave, and the research on them is still in the laboratory stage. Additionally, it is worth mentioning that mature commercial products based on PbS and PbSe bulk materials with a cut-off wavelength of 3–5 μm have been around for a long time. At present, lead salt detectors on the market are mainly single-point detectors with an mm2 detection area level. Lead salt photodetectors, which are competitive in the market, commonly show the detectivity of 10^{10} Jones.

2.1. Physical Properties of PbS

The study on the optical properties of PbS mainly focuses on single or polycrystalline films in the early stage. Many studies have shown differences in light scattering and absorption in nanostructures compared to bulk materials. Therefore, the particle size effect on the optical properties of semiconductors has attracted much attention.

Exciton, formed by direct light excitation or by the combination of free carriers [27], is an important concept in understanding nanoparticle properties. The delocalization region of excitons may be much larger than the semiconductor lattice constant, which may even be several or tens of nanometers. Although nanoparticle sizes are comparable to exciton radii, their physical properties are greatly changed [28]. In the case of PbS nanoparticles, the band gap in the electron spectrum increases from 0.41 to 1.92 eV when the particle diameter is reduced to 5 nm, and the band gap continues increasing when the diameter is further reduced [29]. This indicates that the band gap increases with the decrease in semiconductor particle size, which means a blue shift of the absorption band. There are several reports on optical transmission properties of PbS nanoparticle films, which is summarized in the following section.

2.1.1. A Quantitative Estimation Method for Band Gap Width

To determine the width of the band gap, the most efficient way is to observe the part of the spectrum where the transmittance varies significantly with wavelength. Experimental data show that the most significant dropping part is the region from 700–800 to 2500–2600 nm corresponding to the photon energies from ~1.6 to ~0.5 eV.

Quantitative estimation of semiconductor band gaps by the absorption spectrum [28] depends on the absorption coefficient, which indicates the absorption rate of 1 cm-thick material. The formula is as follows:

$$\sigma = -logT/s \tag{1}$$

where absorption coefficient σ is the optical density associated with a 1 cm-thick layer of material, T stands for the transmittance in relative units, and s is the material thickness, with unit cm.

In order to investigate the relationship between the spectral dependence of the absorption coefficient and the semiconductor band gap, the following formula is introduced [30]:

$$(\sigma(\omega)\hbar\omega)^{1/n} = A_n(\hbar\omega - E_g) \tag{2}$$

where $\sigma(\omega)$ is the spectral dependence of the absorption coefficient; ħ is the reduced Planck constant, which is used to describe quantum size; $\hbar\omega$ is the photon energy; A_n is a coefficient that is dependent on the type of transition and independent of ω; and E_g represents the semiconductor band gap. Then, there is the discussion of the choice of n. The value of n depends on the band gap type: $n = 1/2$ for the direct allowed transition; $n = 3/2$ for the direct forbidden transition; $n = 2$ for the indirect allowed transition; and $n = 3$ for the indirect forbidden transition. If it is based on reflectance spectroscopy, $\sigma(\omega)$ should be replaced by $F(R_\infty)$, which is understood by the following formula:

$$F(R_\infty) = \frac{(1-R)^2}{2R} \tag{3}$$

where R is the reflection coefficient.

When $n = 1/2$ and the electron–hole interaction is ignored, Formula (2) can be simplified as follows:

$$(\sigma(\omega)\hbar\omega)^2 = A^2(\hbar\omega - E_g) \tag{4}$$

Formula (4) is more convenient for data processing and band-gap calculation. However, if the electron–hole interaction cannot be ignored, the spectral dependence of the absorption coefficient makes more sense in the following form [31]:

$$\sigma(\omega) = \frac{B}{1 - exp\left[-C\left(\hbar\omega - E_g\right)^{-1/2}\right]} \tag{5}$$

where B and C are constants. However, in PbS films, due to a high dielectric constant, the electron–hole interaction is not obvious. As a result, Formula (4) is a good choice for the direct allowed transition.

For monodisperse nanoparticles, the absorption spectrum has an easy-to-understand smoothing function with respect to wavelength or energy. However, thin films normally take on the form of polydisperse particles. As a result, Sadovnikov et al. proposed a model to solve this situation [32], which is as follows:

$$\sigma(\omega) = \sum_i c_i \sigma_i(\omega) = \sum_i \frac{c_i A\left(\hbar\omega - E_{g_i}\right)^{1/2}}{\hbar\omega} \tag{6}$$

where c_i is the relative number of nanoparticles of a given size in a film and E_{g_i} is the band gap corresponding to the above nanoparticles. Different grain sizes correspond to different i in PbS nanoparticle films. The final absorption spectrum is the superposition of the corresponding spectra of nanoparticles of different sizes (or different i). Sadovnikov et al. deduced mathematically that the visible band gap E_{g_Σ} is an additive function of the corresponding quantities of nanoparticles of different sizes [32]. By transforming Formula (6), we can obtain

$$[\sigma E]^2 = A^2\left[\sum_i c_i\left(E - E_{g_i}\right)^{1/2}\right]^2 \tag{7}$$

In addition, due to $[\sigma E]^2 = A^2\left(E - E_{g_\Sigma}\right)$, we can simplify Formula (7) to

$$E_{g_\Sigma} \approx \sum_i c_i^2 E_{g_i} + 2\sum_{i \neq j} c_i c_j E_{g_i} E_{g_j}{}^{1/2} + \frac{\left|E_{g_i} - E_{g_j}\right|}{3} \tag{8}$$

When the film has monodisperse nanoparticles $c_1 = 1$ and $c_{i \neq 1} = 0$, in this case, the Formula (8) can be reduced to a boundary condition:

$$E_{g_\Sigma} = E_{g_1} \tag{9}$$

This highly developed interface between the grains of continuous thin films makes the positions of conduction bands and valence bands in the electronic structures of the grains fuzzy, resulting in uncertainties regarding the band gap. The effect of this property on the film spectrum is similar to the difference in the size of nanoparticles [32].

2.1.2. Properties

The photo-response of photodetectors based on PbS thin films has been fully studied by many groups. Here, we provide data measured by Reisfeld et al., who used current-voltage (I–V) characteristics under different lighting intensities, as shown in Figure 2a,b.

Peterson et al. prepared PbS CQDs according to the method of Hines and Scholes [33,34]. They found that these CQDs exhibit distinct exciton peaks in their absorption spectra and band edge luminescence, which can be tuned throughout the NIR region [33] (Figure 2c). To make the measurements using a Si charge-coupled detector (CCD), they synthesized PbS CQDs with the smallest possible particle size, which fluoresced at wavelengths < 1000 nm. The maximum absorption and emission values of the first exciton are 780 and 930 nm, respectively, with a full width at a half-maximum of 170 meV [33] (Figure 2c). Brown et al.

confirmed that the energy band of the CQDs can be changed through ligand exchange, resulting in energy level drifts of the CQDs as high as 0.9 eV [35]. Figure 2d shows the chemical structure of the ligand they used, and Figure 2e shows the complete energy level diagram of PbS CQDs undergoing ligand exchange. Tang et al. studied the effect of the two-step ligand exchange method on the light response of PbS CQDs [36]. The absorption spectra and emission spectra of CQD films prepared by the classical one-step method and the optimized two-step ligand exchange method are shown in Figure 2f. Ushakova et al. studied the control of the process of CQD self-assembly, superlattice uniformity and interparticle distance by optimizing the number of surface ligands. In addition, annealing and aging also affect the structural stability and optical properties of superlattices [37]. Figure 2g illustrates the evolution of small-angle X-ray scattering (SAXS) patterns of superlattices (SLs) formed by 4.3 nm QDs with the time of sample annealing. The interference peak corresponding to the ordered QD ensemble vanishes, and the optical properties of the SLs undergo changes in parallel. In Figure 2h, the absorption and photoluminescence (PL) spectra of SL $QD_{4.3}$ are compared with those of the initial 4.3 nm QD solution. After 1 min of annealing, the absorption and PL bands become broader, and the PL intensity decreases [37].

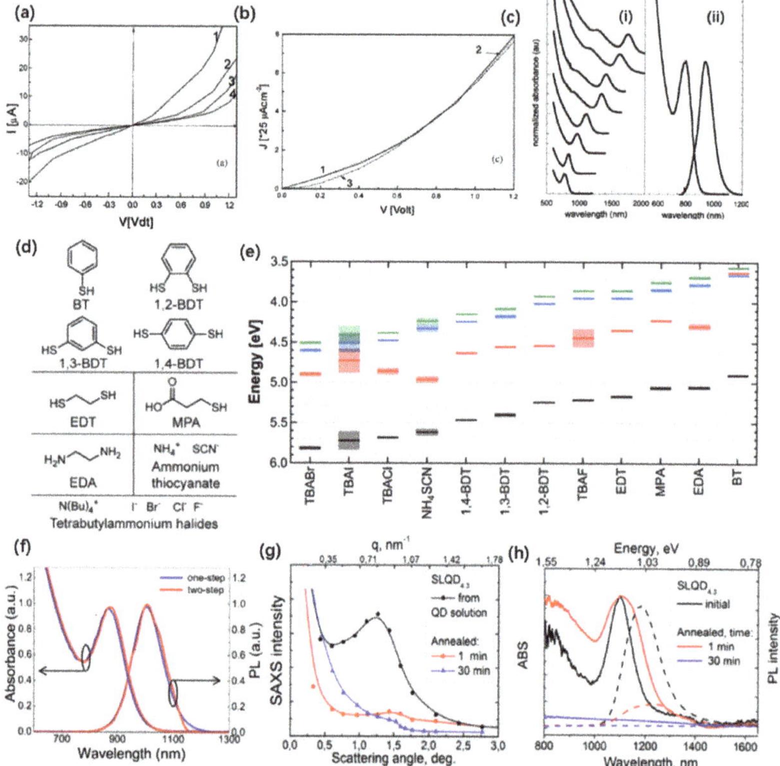

Figure 2. (**a**) I-V characteristics of PbS NCs measured with different NCs mole %. Reprinted with permission from Ref. [38]. Copyright 2002 Springer Nature. (**b**) I-V characteristic of PbS NCs measured at various temperatures. Reprinted with permission from Ref. [38]. Copyright 2002 Springer Nature. (**c**) (i) Absorption spectra of PbS CQDs with different sizes. (ii) Typical absorption (left curve) and fluorescence (right curve) spectra of PbS QDs. Reprinted with permission from Ref. [33]. Copyright 2006 American Chemical Society. (**d**) Chemical structures of the ligands studied by Patrick. Reprinted with permission from Ref. [35]. Copyright 2014 American Chemical Society. (**e**) Energy level diagrams

of PbS CQDs after ligand exchange, which is shown in d. Reprinted with permission from Ref. [35]. Copyright 2014 American Chemical Society. (**f**) The absorption and fluorescence spectra of PbS CQD films were studied by one-step (blue) and two-step (red) ligand exchange methods. Reprinted with permission from Ref. [36]. Copyright 2019 American Chemical Society. (**g**) SAXS pattern of SLs formed by 4.3 nm QDs: initial (black), annealed for 1 min (red) and annealed for 30 min (blue). Reprinted with permission from Ref. [37]. Copyright 2016 American Chemical Society. (**h**) Absorption (ABS) (solid curves) and PL (dashed curves) spectra of $SLQD_{4.3}$: initial (black), annealed for 1 min (red) and annealed for 30 min (blue). Reprinted with permission from Ref. [37]. Copyright 2016 American Chemical Society.

2.2. Fabrication Methods

There have been many mature methods on the preparation of PbS CQDs. For PbS CQDs, there are two major organic synthesis methods [34].

One way is to react lead oleate and bis(trimethylsilyl) sulfide (TMS) in octadecene (ODE). They produce monodisperse CQDs that are 2.6 to 7.2 nanometers in size with corresponding absorption peaks ranging from 825 to 1750 nanometers. However, not all sizes of CQDs made in this way are air-stable [39]. The following are the detailed operation steps: First, the lead oxide (PbO) in oleic acid (OA) is heated under argon or vacuum to prepare lead oleate. After that, a solution of TMS in ODE is injected into the solution until the ratio of lead to sulfur is 2:1. It is worth mentioning that trioctylphospine (TOP) can also be used as a solvent for TMS and has no material effect on the reaction results. The reaction temperature is controlled according to the desired particle size. The NCs are then precipitated with a polar solvent such as methanol or acetone and are subsequently redispersed into an organic solvent such as chloroform or toluene. Precipitation and redissolution are repeated to ensure the removal of the reaction solvents. Finally, the aqueous NC dispersion is centrifuged to remove any remaining impurities [34].

The second method is to react lead chloride ($PbCl_2$) and elemental sulfur in oleylamine (OAm) under nitrogen. PbS NCs were prepared by thermal injection, similar to the first method. The size range (4.2–6.4 nm) of CQDs prepared by this method is much smaller than that of the first method. In addition, the corresponding absorption peak range is from 1200 to 1600 nanometers. The CQD films prepared by this method show good optical stability [40]. In addition, there are many preparation methods of PbS CQDs, most of which are based on the improvement of the above two methods.

Preparing PbS CQD solids is further mentioned in many articles, and different authors differ in method details. Here, we briefly introduce a method offered by McDonald et al. It is necessary to carry out ligand exchange first, and this method is an improvement upon Hines's method [34]. They precipitated the OA-coated NCs with methanol, dried them and dispersed them in excess octylamine. The solution was heated, and the octylamine-coated NCs were precipitated with N, N-dimethylformamide and then redispersed into chloroform. The NCs were then mixed with 2-methoxy-5-(2′-ethylhexyloxy-pphenylenevinylene) (MEH-PPV). The P-phenylenevinylene (PPV) hole transport layer was rotated onto an indium tin oxide (ITO)-coated glass sheet and annealed in vacuum to allow polymerization. A mixture of MEH-PPV and NCs was dissolved in chloroform and spun onto the PPV layer to create a thin film. Finally, the upper contact was prepared by vacuum evaporation [41].

2.3. Devices

PbS NC films have been used in many devices, including infrared LEDs [42], mid- and long-wave infrared detectors [43], upconversion photodetectors [44] and field effect transistors (FET) [45]. Here are some typical concrete examples.

A hybrid graphene PbS QD phototransistor with ultra-high gain has been reported (Figure 3a). Using the strong light absorption of QDs and the high mobility of graphene, the materials were mixed into a system to make it have high sensitivity for light detection. The ultra-high gain of graphene phototransistors was realized for the first time by using the charge transfer of the two materials [46]. Coupling with graphene has always been a research direction for the improvement of the performance of detectors.

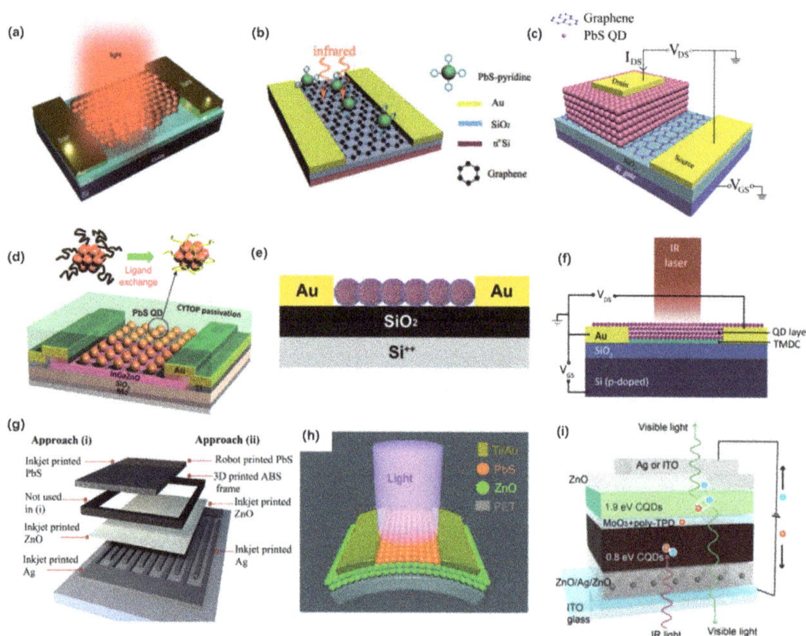

Figure 3. (**a**) Schematic diagram of graphene QD hybrid phototransistor; graphene sheets are deposited on the SiO_2/Si structure coated with PbS QDs. Reprinted with permission from Ref. [46]. Copyright 2012 Springer Nature. (**b**) Schematic diagram of a graphene transistor modified with PbS QDs under illumination. Reprinted with permission from Ref. [47]. Copyright 2012 John Wiley and Sons. (**c**) The device architecture schematic of the graphene semiconductor metal sandwich construction. Reprinted with permission from Ref. [48]. Copyright 2016 AIP Publishing. (**d**) A schematic 3D view of the ligand exchange process and the PbS QD/IGZO hybrid phototransistor. Reprinted with permission from Ref. [49]. Copyright 2016 Springer Nature. (**e**) PbS CQD FETs based on blade-coated films. Reprinted with permission from Ref. [45]. Copyright 2018 American Chemical Society. (**f**) Hybrid photodetector architecture with circuit diagram. Reprinted with permission from Ref. [50]. Copyright 2019 American Chemical Society. (**g**) Sketch of the photo-conducting devices prepared by two alternative all-printing approaches. Reprinted with permission from Ref. [51]. Copyright 2019 American Chemical Society. (**h**) Schematic diagram of the two-terminal contact flexible photodetector. Reprinted with permission from Ref. [52]. Copyright 2018 Springer Nature. (**i**) Schematic diagram of the structure which uses CQDs as the infrared sensitive layer and CdSe/ZnS core shell QDs as top visible light emission layer. Reprinted with permission from Ref. [44]. Copyright 2020 Springer Nature.

Here is another PbS QD photodetector coupled with CVD-grown graphene (Figure 3b). It shows amazing responsivity of 1×10^7 AW^{-1} at a power of 30 pW [47]. Graphene was also used in PbS QD field effect phototransistors as electrodes (Figure 3c). PbS QDs are used as channels, and field-effect transistors show good responsivity of 4.2×10^2 A/W [48]. By comparing specific the detectivity of these PbS–graphene devices, the device reported by Konstantatos et al. shows the best property which reaches a specific detectivity of 7×10^{13} Jones.

Besides graphene, many new materials are candidates for coupling materials of PbS devices. A PbS QD-sensitized InGaZnO photoinverter for NIR detection is reported in 2016 (Figure 3d). This hybrid photoelectric device has good light-response performance of high specific detectivity of 10^{13} Jones [49], which is similar to that of the best performing optoelectronic devices based on PbS–graphene, as mentioned before.

In addition, in the research of lead sulfide detectors, there are many novel methods to treat thin films. Single-step fabrication of FET using CQD ink was studied, and the authors of this study also reported the importance of removing ligands after deposition. The structure of the FETs based on blade-coated PbS CQD films is as shown in Figure 3e. They showed that this superficial coating method with blades can prepare high-quality thin films, and the device has a good response [45]. It is also the work of many researchers to make devices by coupling other materials with PbS to expand spectral range. A hybrid PbS CQD transition metal dichalcogenides photodetector with high sensitivity was reported in 2019 (Figure 3f). The authors of this study extended the spectral coverage of this technology from 1.5 μm to 2 μm and showed excellent specific detectivity of 10^{12} Jones at room temperature [46].

Perovskite is a hot material that has been used with 3D-printing technology in recent years. As a result, IR photodetectors made of PbS nanotubes with perovskite ligands were studied and reported in 2019 (Figure 3g). This all-printed device shows a cut-off frequency of over 3 kHz and a high detectivity of 10^{12} Jones [51], which shows the same order of magnitude as the hybrid PbS CQD transition metal dichalcogenides photodetector, as mentioned before.

The fabrication potential of flexible devices is recognized as a great advantage of CQD materials. A flexible broadband photodetector based on the heterostructure of PbS/ZnO nanoparticles was reported in 2019 (Figure 3h), and it shows the widest detectable spectral range (UV-Vis to NIR) of devices that we introduce in this section [52]. This detector shows a high detectivity of 3.98×10^{12} Jones.

Upconversion device is an exciting field because of its unique properties. A QD-based solution-treated upconversion photodetector was studied and reported (Figure 3i). The photodetector has a low dark current, a high detectivity of 6.4×10^{12} Jones, a millisecond response time and compatibility with flexible substrates [44].

3. PbSe

PbSe thin films are widely used for NIR and MWIR range applications due to their unique physical properties. PbSe polycrystalline thin films are widely used for infrared detectors [53]. Considering that the photosensitivity of thin films is very sensitive to crystallite size, a research group prepared thin films with a thickness of 1.2 μm and different crystallite sizes. After post-processing, the detectivity of the photodetector achieved 2.8×10^{10} Jones at room temperature [54]. PbSe has many forms, such as polycrystal, monocrystal and QDs, which have been studied deeply and are widely used in infrared detectors. In order to reduce costs and improve performance, there has been a lot of research on nanotechnology in the past few decades [20]. Photonic applications of CQDs involving lead chalcogenides are mainly associated with PbS and PbSe.

3.1. Properties

As lead chalcogenides, PbS and PbSe have many similarities in optical properties. PbSe is a typical direct band-gap semiconductor with a narrow band gap of 0.27 eV at room temperature [55]. The narrow band property of bulk PbSe makes it ideal for MWIR detection. The small electron effective mass in PbSe causes a large Bohr radius of 46 nm, which makes the material ideal for studying quantum size effects observed only in large particles with a small surface-to-volume ratio [56]. Combining these properties makes it possible to precisely alter the band gap and the spectral range of optical photoresponsivity [20].

Using sodium selenite sulfate as the selenium source and lead acetate as the lead source, Begum et al. prepared nanocrystalline PbSe films on glass substrate by the chemical bath deposition (CBD) method. Its optical properties can be referred to in Figure 4a. The results show that the optical absorption of PbSe films increases with increases in deposition temperature. This may be due to an increase in grain size and a decrease in defects [57]. The $(\alpha h\nu)^2$ vs $(h\nu)$ plots of PbSe thin films are linear over a wide spectral range, as shown in Figure 4b. This indicates that there is a direct optical band gap in the PbSe films [58]. Zhu et al. used pulse the sonoelectrochemical synthesis method to prepare PbSe nanoparticles and estimated the band gap of the materials by optical diffuse reflectance spectroscopy [59]. The mathematical basis they used to estimate the band gap width can be referred to in Section 2.1.1. The band gap of PbSe prepared by this method was 1.10 eV. The normalized $(F(R)\times h\nu)^2$ vs $h\nu$ (eV) of PbSe nanoparticles can be referred to in Figure 4b.

The absorption spectra of PbSe CQDs with different particle sizes were given by Gao et al. (Figure 4d) [60]. They used PbSe CQDs with an average radius of 2.8–3.5 nm and a corresponding band gap of 0.60–0.76 eV, judging from the position of the band edge peak in the absorption spectrum. Thambidurai et al. developed a high-performance infrared photoelectric detector up to 2.8 μm based on PbSe CQDs [61]. They gave the photocurrent and voltage characteristics of photodetectors based on PbSe CQDs with different thicknesses (500, 900 and 1400 nm) in the wavelength range of 1.5–2.8 μm [61]. The photoelectric current and voltage characteristics of PbSe CQDs photodetectors with three different thicknesses and different LED illumination are shown in Figure 4e–h.

Ahmad et al. studied PbSe CQD solar cells with >10% efficiency. To explore energy disorder in both CQD films, Urbach behavior was investigated by tail-state absorption, as shown in Figure 4i. Sharp band tails in a range of 1.0 to 1.25 eV were observed for lead iodide (PbI_2)-capped PbSe CQD films deposited via the one-step technique, and they exhibited lower Urbach energy (28 meV). However, Urbach energy in PbI_2-treated and tetrabutylammonium iodide (TBAI)-treated PbSe CQD layer-by-layer (LBL) films were 43 and 83 meV, respectively [62].

3.2. Fabrication Methods

Photodetectors based on PbSe can be divided into two categories depending on the target wavelength of photosensitivity. In the early days, PbSe intrinsic semiconductors were widely used in MWIR detectors. Bulk PbSe has an optical band gap of 0.27 eV and a sensitive wavelength of 4.4 μm. In recent decades, however, there has been much more discussion about PbSe low-dimensional materials. Scholars have turned their attention to exploiting the quantum confinement effect, referring to the phenomenon that the energy quantization of microscopic particles becomes more obvious with decreases in its space motion limitation size, and the energy level changes from a continuous energy band to a discrete energy level, especially when the ground state energy level moves up and blue shift occurs. A common idea is to reduce the grain size below the Bohr radius. The sensitivity wavelength can be as short as 690 nm, and the band gap can reach 0.18 eV by adjusting the grain size of PbSe [20]. PbSe-based infrared detectors have the potential to span the MWIR, SWIR, NIR and even visible wavelengths.

Massive PbSe semiconductor films prepared by chemical water baths and progress in preparation of PbSe NCs are described here.

3.2.1. Chemical Bath Deposition

CBD is a simple and effective method to synthesize high-quality semiconductor thin films without expensive and complex equipment [63]. However, it should be noted that film properties with CBD are greatly affected by precursor fluid composition, bath time, bath temperature and PH value. CBD films are often considered to be deposited on silicon, glass or gallium arsenide (GaAs) substrates with a thickness between 0.2 and 2 μm [20,54,64]. Some scholars believe that a rough substrate surface leads to better deposition [65].

Regarding CBD, there are many different combinations of precursors. In 2003, Hancare et al. prepared PbSe films using lead nitrate and selenosulphate as precursors and tartaric acid as the complexing agent. PbSe thin films were deposited onto cleaned, spectroscopic-grade glass substrates [66]. In 2010, Kassim et al. prepared PbSe films with lead nitrate solution as the lead source, sodium selenate as the selenium source and tartaric acid as the complexing agent [67]. In 2013, Qiu et al. carried out research on the room-temperature PbSe photodetector, mixing sodium hydroxide, lead acetate and selenosulfate into a precursor solution in a ratio of 12:1:1 [54].

In addition, there are few references about the influence of the timing of the chemical water bath deposition process. Anuar et al. found that the duration of the water bath affects grain size, film thickness and surface roughness. They demonstrated in reliable experiments that grain size, film thickness and surface roughness increase when the deposition time increases from 20 to 150 min. The atomic force microscopy (AFM) images of different samples indicate that the film obtained by 60 min of deposition was uniform and that the substrate surface was covered with good spherical particles [68]. Hone et al. prepared three different samples with deposition times of 30, 45 and 60 min. According to X-ray diffraction analysis (XRD) patterns, they found that the peak intensities and preferred orientations of the crystals were affected by different deposition times. When choosing deposition times of 30 and 60 min, the crystals preferred orientations along the (111) plane, whereas for a deposition time of 40 min, the crystals preferred an orientation along the (200) plane [69].

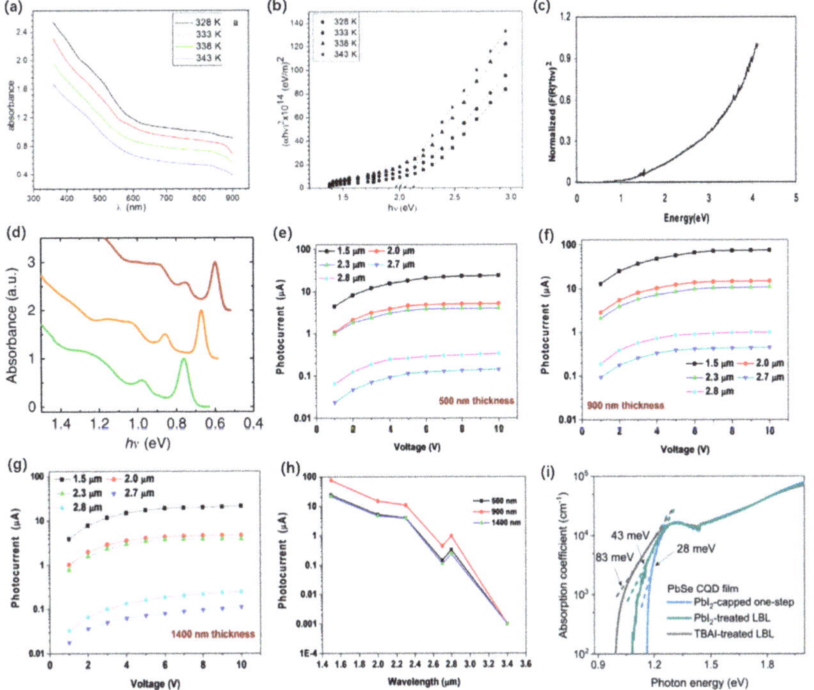

Figure 4. (**a**) UV absorption spectra of PbSe thin films reported by Anayara et al. Reprinted with permission from Ref. [57]. Copyright 2012 Beilstein-Institut. (**b**) $(\alpha h\nu)^2$ vs. $(h\nu)$ plots of PbSe reported by Anayara. Reprinted with permission from Ref. [57]. Copyright 2012 Beilstein-Institut. (**c**) Normalized $(F(R) \times h\nu)^2$ vs. $h\nu$ (eV) of PbSe by pulse sonoelectrochemical synthesis method. Reprinted with permission from Ref. [59]. Copyright 2000 American Chemical Society. (**d**) Absorption spectra of three QDs with different radii of 2.8 nm (green), 3.0 nm (orange) and 3.5 nm (red). Reprinted with permission from Ref. [60]. Copyright 2015 Springer Nature. (**e–g**) I-V characteristics of devices under different illuminations with wavelengths from 1.5 µm to 2.8 µm for PbSe thin films with different thicknesses. Reprinted with permission from Ref. [61]. Copyright 2017 Optical Society of America. (**h**) The photocurrent of the devices under different illuminations with wavelengths from 1.5 µm to 3.4 µm. Reprinted with permission from Ref. [61]. Copyright 2017 Optical Society of America. (**i**) Urbach energy of three differently treated films. Reprinted with permission from Ref. [62]. Copyright 2019 John Wiley and Sons.

The influence of temperature and PH value on CBD has been widely reported. Deposition temperature has always been considered the most important parameter affecting film quality. It is believed that, with increases in deposition temperature, grain size increases, and dislocation density and microstrain decrease. Reductions in dislocation density and microstrain indicate reductions in lattice defects, i.e., the improvement of film quality. Additionally, deposition temperature has a significant effect on preferred orientation and film thickness [70].

CBD polycrystalline films have good properties when sensitized with oxygen and iodine, which is necessary to activate PbSe as an MWIR detector. However, the specific mechanism of sensitization has not been clearly defined. The sensitization process of PbSe varies from reference to reference, but it usually involves two thermal steps: oxidation and iodization. In order to obtain better performance for PbSe infrared detectors, it is very important to study the mechanism behind the sensitization process, but there are still many doubts.

CBD also shows good prospects in the preparation of PbSe NCs. The chemical characteristics of PbSe thin films are strongly influenced by growth conditions such as ion concentration, PH value and deposition time. Studies have shown that the average size of PbSe nanoparticles increases from 23 nm to 51 nm as the deposition time passes from 1 h to 16 h [71]. In addition, increases in temperature also make the grain size larger [72].

In addition to chemical water bath deposition, PbSe films can also be synthesized by a variety of deposition techniques, such as co-evaporation [73], pulse acoustic electrochemical [59], thermal evaporation [74] and pulse laser deposition [75]. Figure 5 shows the schematic diagrams of CBD and the CQD synthetic procedures.

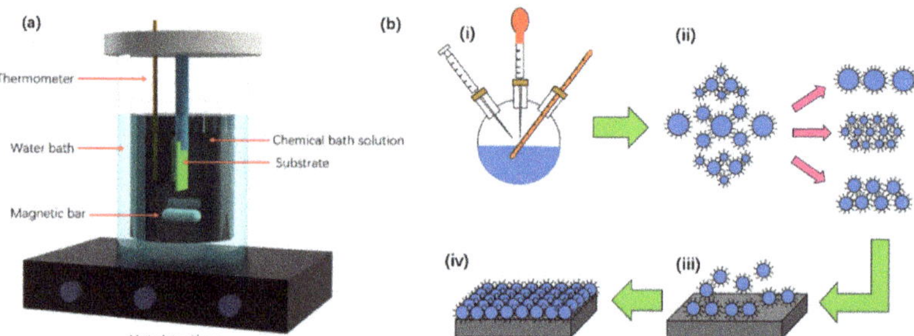

Figure 5. (**a**) Schematic diagram of chemical bath deposition. (**b**) Schematic diagram of the CQD synthetic procedures: (i) Synthesize CQDs by solution-phase routes; (ii) Reduce the size distribution of CQD samples by size-selective precipitation; (iii) Deposit CQD dispersions that self-assemble; (iv) Form ordered CQD assemblies.

3.2.2. Fabrication of PbSe NCs

Monodispersion is required for photodetectors based on PbSe CQDs. A rapid nucleation followed by a slow growth process is considered the key [76]. Nucleation is affected by temperature, degree of supersaturation in solution, interfacial tension, etc. [77]. There are two ways to stop nucleation. One way is to lower the concentration of the solution below a certain level. The other is to rapidly inject precursors into a high-temperature mixed solution, which has achieved the purpose of rapid cooling, commonly known as thermal injection [78]. Thermal injection is considered to be the most widely used method for synthesizing CQDs.

Murray et al. reported a method using lead oleate as the lead source and trioctylphosphine selenide as the selenium source, and the two are dissolved in trioctylphosphine. The above-room-temperature solution is quickly injected into a fast-stirring solution containing diphenylether at 150 °C. The growth rate of NC can be accelerated by increasing the solution temperature, and NC with the larger size can be prepared at a higher temperature. Solution temperatures of 90–220 °C correspond to NC diameters of 3.5–15 nm. When the grains reach the target size, the dispersion is cooled, short-chain alcohols are added to flocculate the NCs, and it is then separated from the solution by centrifugation [79].

3.3. Devices

Research on PbSe photodetectors is focused on improving efficiency, making large imaging FPAs, manufacturing thermoelectric cooling imaging systems, and making more compact and low-cost systems [20]. There are many types of photodetectors based on PbSe, such as photoconductor [80], phototransistor [81] and photodiode [81].

Regarding the most advanced PbSe photodetector equipment, there are mainly the following kinds: broadband photodetectors using PbSe QD [82], PbSe-based photodiode detectors [83], tandem photodiode detectors [84] and PbSe-based FET detectors [85].

Jiang et al. reported an ultra-sensitive tandem CQD photodetector (Figure 6a,b), which shows maximum detectivites of 8.1×10^{13} Jones at 1100 nm and 100 K [83].

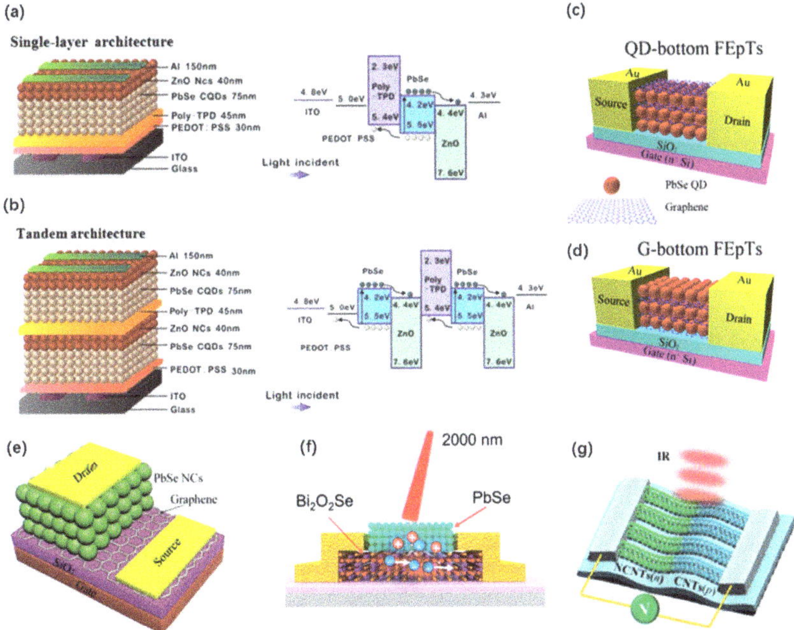

Figure 6. (**a**,**b**) Device structures and band diagrams of single-layer and tandem photodetectors. Reprinted with permission from Ref. [83]. Copyright 2015 RSC Pub. (**c**,**d**) Illustrations of QD-bottom FEpTs (**c**) and G-bottom FEpTs (**d**). Reprinted with permission from Ref. [86]. Copyright 2015 American Chemical Society. (**e**) Schematic illustration of PbSe NC-based VPT. Reprinted with permission from Ref. [85]. Copyright 2019 Elsevier. (**f**) Schematic illustration of the PbSe CQD–Bi_2O_2Se nanosheet hybrid photodetector. Reprinted with permission from Ref. [87]. Copyright 2019 American Chemical Society. (**g**) Schematic illustration of the detector's structure. Reprinted with permission from Ref. [80]. Copyright 2016 RSC Pub.

Many research groups have tried to prepare IR photodetectors by coupling graphene with PbSe CQD. A heterojunction phototransistor based on PbSe CQD–graphene hybrids (Figure 6c,d), which shows the highest responsivity of 10^6 A/W, was reported in 2015 [86]. In addition, the preparation of graphene electrodes on PbSe CQD vertical phototransistors has also been tried (Figure 6e). The phototransistor exhibits an excellent responsivity of 1.1×10^4 A W^{-1}, a detectivity of 1.3×10^{10} Jones, and an external quantum efficiency of 1.7×10^6% [85]. Besides graphene, coupling between other materials and PbSe materials has also been studied and published. A hybrid photodetector based on Bi_2O_2Se nanosheets sensitized by PbSe CQDs was reported (Figure 6f). Compared to pure Bi_2O_2Se or PbSe CQDs, the interfacial band offset between the two materials enhances the device's responsivity and the response time. This PbSe CQDs–Bi_2O_2Se photodetector can render an infrared response above 10^3 A/W at 2 μm under external field effects [87].

In addition to coupling a novel two-dimensional material with PbSe QD to obtain a special photodetector, flexible devices are an important development direction of this material. Figure 6g is a schematic diagram of a flexible device which puts n- and p-type carbon nanotubes (CNTs) on a flexible substrate. The IR response of the photodetector is derived from the interface between the n- and p-type CNTs and the Thomson potential. By the CVD method, two kinds of CNT arrays are grown on the substrate and construct the CNT p–n junction [80].

4. HgTe

In recent years, HgTe CQDs, as a new choice of infrared detectors, have attracted much attention because of their excellent optical properties, processability and adjustable absorption characteristics. However, at present, reported HgTe CQD infrared detectors mainly focus on short and mid-wave IR photodetectors. Additionally, research progress on LWIR is slow, and there is still a gap between that and commercial detectors. Although the research on this material is still in the laboratory stage at present, its potential in preparation cost, response speed and coupling with flexible substrates is still exciting.

4.1. Properties

Keuleyan et al. firstly synthesized MWIR HgTe CQDs [88] and successfully applied them in MWIR photodetection [89]. They prepared the HgTe CQDs by a simple two-step method and found that the absorption, photodetection with sharp edges, and narrow photoluminescence are tunable between 1.3 and 5 μm [88]. Herein, we choose to display the absorption spectra of HgTe CQDs' solutions in C_2Cl_2, which shows the relationship between absorption characteristics and sample size (as shown in Figure 7d). Keuleyan et al. found that larger particles give an absorption onset at lower frequencies.

Keuleyan et al. also systematically studied the electronic structure and size-dependent spectrum of HgTe CQDs, and they extended the HgTe CQD spectra covering LWIR [90]. Figure 7a shows the successful extension of the spectrum up to 12 μm at 80 K, which includes the LWIR. Figure 7b shows the absorption spectra of the corresponding films at room temperature.

Allan and Delerue theoretically studied the electronic structure and energy gap of HgTe CQDs [91]. The energy gap of spherical HgTe CQDs is plotted versus size in Figure 7c. It varies according to the form of $1/\left(0.02126 \times d^2 + 0.21562 \times d + 0.01684\right)$ (in electrovolts), where d is the diameter in nanometers [91]. In addition, they came to a conclusion that the gaps of QDs with cubic, tetrahedral or octahedral shapes are almost the same as the gaps of spherical QDs with the same size, according to previous research [92,93].

However, partial aggregations in HgTe CQD always appear in Keuleyan et al.'s methods, causing difficulties in surface modification for those CQDs. This challenge was solved in 2017, when Shen et al. developed a new recipe for nonaggregating HgTe CQDs, which can be stabilized without any thiols [94].

Later, in 2018, Hudson et al. studied the conduction band fine structure of HgTe CQDs. They synthesized highly monodispersed HgTe CQDs and tuned their doping both chemically and electrochemically [95]. Splitting of the intraband peaks was observed corresponding to nondegenerate $1P_e$ states because of the size uniformity of the CQDs and because of strong spin-orbit coupling in HgTe [95]. We borrow the highlighted picture from the article, as shown in Figure 7e.

Later, in 2019, Chen et al. achieved high carrier mobility in HgTe CQD solids with a hybrid ligand exchange method and improved the detectivity in mid-IR photodetectors [96]. Herein, we show detectivity from 80 to 300 K using measured responsivity and noise at each temperature, as shown in Figure 7f. The 120 nm-thick HgTe/hybrid device shows a higher detectivity than the 260 nm thick HgTe/EDT device at all temperatures. It is indicated that the HgTe/hybrid ligand material is 10 times better than HgTe/EDT at 80 K and 5 better times at 300 K. Since higher mobility allows longer carrier diffusion lengths and higher operation temperatures in the geminate recombination regime, the transport improvement generally benefits CQD photodetection devices [96].

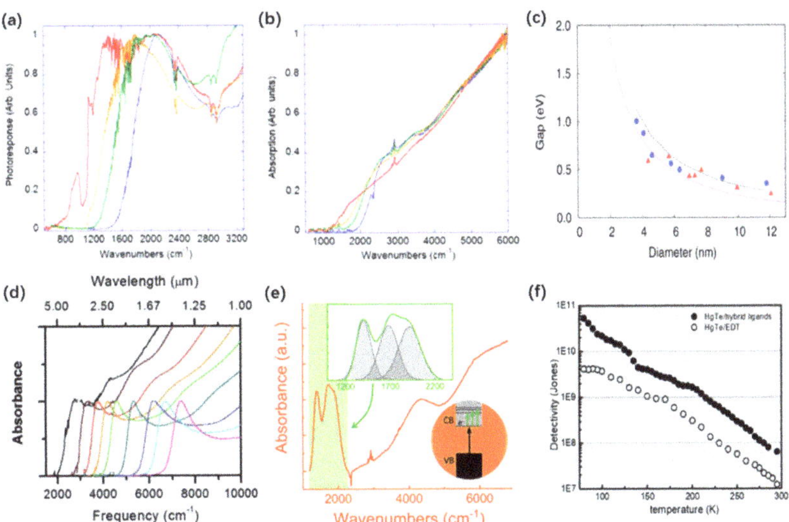

Figure 7. (**a**) Photo-response of thin films of HgTe QDs on a Si/SiO_2 substrate. Samples were made with different temperatures of 100 °C (blue), 110 °C (green) and 120 °C (orange). The red-line sample was obtained with a regrowth. Reprinted with permission from Ref. [90]. Copyright 2014 American Chemical Society. (**b**) Absorption spectra of thin films of HgTe QDs with the same color scheme as in (**a**). Reprinted with permission from Ref. [90]. Copyright 2014 American Chemical Society. (**c**) Energy gap of spherical HgTe QDs (300 K) versus diameter (magenta solid line) compared to the experimental data of Ref. [97] (red triangles) and Ref. [98] (blue disks). The black-dotted line represents the energy of the second peak in the calculated absorption spectrum. Reprinted with permission from Ref. [91]. Copyright 2012 American Physical Society. (**d**) Absorption of solutions of HgTe CQDs in C_2Cl_4. Reprinted with permission from Ref. [88]. Copyright 2011 American Chemical Society. (**e**) Spectrum of N-doped HgTe QDs indicates that doping leads to intraband absorbance and restrains the first two excitonic transitions. The green box illustration shows that the total absorption spectrum profile is composed of three small overlapping absorption spectra. The illustration with the orange background shows intra and interband transitions. Reprinted with permission from Ref. [95]. Copyright 2018 American Chemical Society. (**f**) Detectivity as a function of temperature for the HgTe/hybrid and HgTe/EDT films. Reprinted with permission from Ref. [96]. Copyright 2019 American Chemical Society.

4.2. Fabrication Methods

HgTe is one of the most difficult nanoparticle compounds to prepare. In order to study the preparation method of II-VI group quantum wells (QWs) and QDs, Brennan et al. focused on the synthesis of the precursors of the target material. They reported the preparation of the solid-state compounds ZnS, ZnSe, CdS, CdSe, CdTe and HgTe from the corresponding $M(ER)_2$ compounds (M = Zn, Cd, Hg; E = S, Se, Te; R = n-butyl, phenyl) and/or phosphine complexes thereof. The idea is to mix the bidentate phosphine 1,2-bis(diethylphosphino)ethane (DEPE) and the $M(ER)_2$ nucleus at a ratio of 1:2 to produce a coordination polymer or a dimeric compound, depending on the material. They found that thermal decomposition of solid compounds produces bulk solid products, whereas thermal decomposition of liquid compounds produces nanoparticle-sized solid products. The team made HgTe by processing $Hg(TeBu)_2$ through photodecomposition [99].

Rogach et al. studied the preparation of HgTe NCs by wet chemical synthesis [100]. There are two main systems for wet chemical synthesis of cadmium chalcogenide NCs. One system involves coating NCs with trioctyl phosphine (TOP) / trioctyl phosphine oxide (TOPO) in a non-aqueous solution [101]. The other system uses various thiols as a stabilizer in an aqueous solution [102–104]. Rogach's method is to pass hydrogen telluride gas buffered with nitrogen through aqueous N_2-saturated mercury(II) perchlorate solutions at a pH of 11.2 with the presence of 1-thioglycerol as an effective size-regulating and stabilizing agent. HgTe CQDs prepared in this way can be precipitated as powders using 2-propanol and then can be capped with thioglycerol on the surface for the purpose of being able to be re-dissolvable in water [100].

Inspired by Huang's work [105], Green et al. reacted mercury(II) bromide and tri-n-octylphosphine telluride to form the intermediate $(HgBr_2)_4{\cdot}(TeP(C_8H_{17})_3)_3$ and eventually the nanosized HgTe, and the intermediates have not been rigorously tested due to their rapid reaction [106].

In 2011, Guyot-Sionnest et al. reported the first synthesis of HgTe CQDs with absorption and emission in the MWIR ranges (>3 μm) [88,89]. They tried to react mercury chloride ($HgCl_2$) dissolved in OAm with telluride (Te) in TOP, as well as mercury(II) acetate with Te dissolved in tri-n-octylphosphine (TOPTe) in alcohol. In 2014, they further extended the absorption edge of HgTe CQDs into LWIR by making a small modification of the growth dynamics with diluted TOP (Te precursor by OAm) [90].

To control CQD growth in an organic solvent, the particles need good surface passivation provided by long chain ligands and slow particle spread to avoid aggregation. In addition, a low temperature is necessary for the synthesis of HgTe CQDs to refrain from growing to bulk sizes. Keuleyan et al. [88] developed methods from Cademartiri et al. [107] for the above two reasons. They formed a viscous mixture with Hg/OAm ratios of about 1:120. Then, they rapidly injected TOPTe to produce Te. Particle size was controlled by the temperature and duration of reaction, with the smallest particles obtained at 60 °C and with the largest obtained at above 100 °C [88].

In 2017, Shen et al. used $HgCl_2$ as the mercury source, bis-(trimethylsilyl) telluride (TMS_2Te) as the tellurium source and OAm as the solvent. They used anhydrous tetrachloroethylene (TCE) to quickly cool the reaction. This reaction can be easily scaled up to 1 mmol HgTe by using 2 mmol $HgCl_2$ and 1 mmol TMS_2Te. The grain size of the HgTe CQDs can be expanded to 11 nm by adjusting the $HgCl_2$ to TMS_2Te molar ratio to 4:1 [95].

4.3. Devices

HgTe CQDs IR photodetectors mainly include three categories: photoconductor, phototransistor and photovoltaic devices [5]. Photoconductors have the simplest device structure and can be prepared by the deposition of CQDs on interdigitated electrodes [108]. Compared with photoconductors, photovoltaic devices, which theoretically avoid 1/f noise and dark currents, can improve the sensitivity of the device [5].

Guyot-Sionnest et al. conducted a lot of relative research [88,89,94–97,109–116] and developed the first background-limited infrared photodetection (BLIP) HgTe CQD MWIR photodetector with a specific detection rate of 4.2×10^{10} Jones, with a response time in microseconds and with coverage of all infrared spectral regions [111]. In 2018, this team further improved the sensitivity and response speed of photodetectors through new doping methods [117]. Diagrams of these devices are shown in Figure 8.

In 2019, Tang et al. reported the first HgTe CQD flexible SWIR and MWIR dual-band photodetector with a high detectivity D* of 7.5×10^{10} Jones at room temperature and a fast response of 260 ns [118]. Integrating the Fabry–Perot cavity with detectors enhances light absorption and the photo-response, with controllable spectral features. (Figure 8d,e). Later, Tang et al. reported a dual-band infrared detector using stacked CQD photodiodes, which provide a bias-switchable spectral response in two distinct bands [11]. It consists of one

SWIR and one MWIR photodiode arranged in an n–p–n structure, with Bi_2Se_3 and Ag_2Te as the n and p layers. By controlling the bias polarity and magnitude, it can be rapidly switched between SWIR and MWIR at modulation frequencies of up to 100 kHz with a D* above 10^{10} jones at cryogenic temperature.

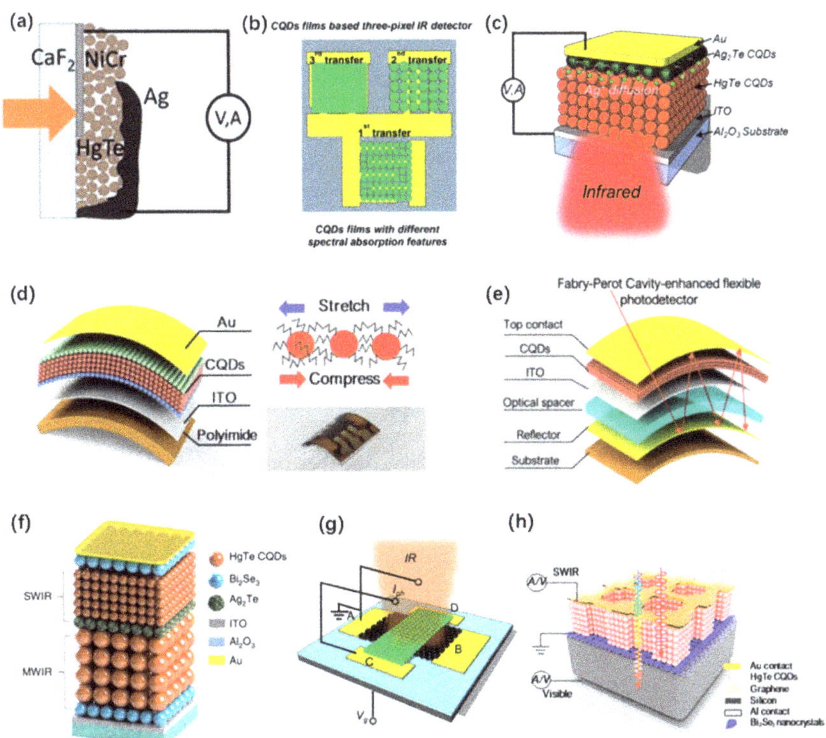

Figure 8. (**a**) Schematic diagram of the device structure. Reprinted with permission from Ref. [111]. Copyright 2015 AIP Publishing. (**b**) Schematic configuration of the three-pixel photodetector. Reprinted with permission from Ref. [108]. Copyright 2016 American Chemical Society. (**c**) Photodetector design. Reprinted with permission from Ref. [117]. Copyright 2018 American Chemical Society. (**d**) Schematic diagram of the device architecture of flexible HgTe CQD photovoltaic detectors. Reprinted with permission from Ref. [118]. Copyright 2019 John Wiley and Sons. (**e**) Schematic diagram of Fabry–Perot cavity-enhanced HgTe CQD detectors. Reprinted with permission from Ref. [118]. Copyright 2019 John Wiley and Sons. (**f**) Schematic diagram of the structure of a dual-band CQD imaging device. Reprinted with permission from Ref. [11]. Copyright 2019 Springer Nature. (**g**) Schematic of the graphene/HgTe CQD junction. Reprinted with permission from Ref. [119]. Copyright 2019 American Chemical Society. (**h**) A dual-channel photodetector made by depositing a CQD infrared photodiode onto a graphene/p-silicon Schottky junction. Reprinted with permission from Ref. [120]. Copyright 2020 American Chemical Society.

In addition, Tang et al. reported a graphene/HgTe QD photodetector with gate-tunable IR photo-response [119]. The graphene/HgTe QD junction combines the high carrier mobility of graphene and tunable infrared optical absorption of HgTe CQDs, which offers a promising route for the next generation of infrared optoelectronics.

In 2020, Tang et al. also reported a visible and infrared dual-channel photodetector. They deposited a CQD IR photodiode onto a graphene/p Si Schottky junction and obtained a responsivity of ~0.9 A/W in the visible spectrum, and the infrared CQD photodiode had a detectivity of $\sim 5 \times 10^9$ Jones at 2.4 μm [120].

5. One- and Two-Dimensional Materials

In this section, we shortly introduce one- and two-dimensional materials used for room-temperature infrared photodetectors. Regarding 1D materials, NWs, especially InAs NWs, are of great concerned due to their high electron mobility [121], easy-to-form ohmic metal contact [122], high electron saturation rate and tiny diameter, which make them mechanically flexible [123]. It has been reported that nanorods are annealed to improve their crystal structures and electrical properties. In addition, detectivity can reach 10^{13} Jones, and the noise equivalent power can reach 10^{-14} W [124]. Regarding this material, we do not give more details. In Section 5, we give more introductions of 2D materials, such as graphene and black phosphorus. Moreover, 2D materials have attracted much attention because of their flexibility, broadband absorption and high carrier mobility. However, because of its nanometer-order thickness, 2D materials have weak absorption, which limits the detection performance of IR photodetectors based on 2D materials [22].

5.1. Properties

5.1.1. Graphene Properties

It is generally believed that a single atomic plane is a 2D crystal, whereas 100 layers are a film of three-dimensional (3D) materials [125]. However, graphene is a special case because its electronic structure evolves rapidly as its number of layers increases, reaching the 3D limit at 10 layers [126]. Only graphene and its bilayers have simple electron spectra, as they are zero-gap semiconductors with one electron and one hole. For layers 3 and above, the spectrum becomes increasingly complex with several carriers present [127], and the conduction bands and the valence bands begin to overlap significantly [126,128].

Peng et al. believe that the size of graphene quantum dots (GQDs) determines its band gap, which ultimately leads to different luminescence conditions when the size of GQDs increases. They obtained three GQDs in different size ranges (1–4 nm, 4–8 nm and 7–11 nm) by controlling the temperature. The resultant GQDs exhibited versatile PL emission color from blue and green to yellow, as the energy gap decreases from 3.90 to 2.89 eV [129].

Since GQDs are fragments derived from graphene flakes, the GQD layer contributes significantly to vertical size, thus altering luminescence properties [130]. Using carbon black as the carbon source, monolayer and multilayer GQDs were prepared simultaneously in nitric acid medium by the top-down method [131].

5.1.2. Black Phosphorus Properties

The band gap of black phosphorus is the most important parameter for determining its light absorption. Band structures characterize the optical properties of 2D materials, especially those that interact with light [132].

Unlike graphene, black phosphorus is formed from a folded honeycomb lattice, which results in strong anisotropic electron mobility and anisotropic optical responses [133]. The layered structure of black phosphorus can be obtained from bulk crystals by mechanical stripping. The most attractive property of black phosphorus is that its layers depend on direct band gaps. The band gap from ~2.0 eV for monolayer, ~1.3 eV for double layers and ~0.8 eV for triple layers lowers down to 0.3 eV for the bulk state [132]. Figure 9c shows the crystal and electronic structure of bulk black phosphorus [134]. Figure 9d,e exhibit the linear absorption and reflection spectra for a few layers of black phosphorus [135,136].

5.2. Fabrication Methods

5.2.1. Graphene Preparation

In recent years, due to the great prospect of graphene, there have been many reports on the preparation methods of graphene films.

There are currently seven reliable methods for the preparation of graphene sheets. They are (i) mechanical cleavage of highly oriented pyrolytic graphite (HOPG); (ii) epitaxial growth on an insulator surface; (iii) chemical vapor deposition (CVD) on the surfaces of single crystals of metals; (iv) arc discharge of graphite under suitable situations; (v) use of intercalated graphite as the starting material; (vi) preparation of appropriate colloidal suspensions in selected solvents; and (vii) the reduction of graphene oxide sheets [137]. The above methods are described in detail in Section 5.2.1.

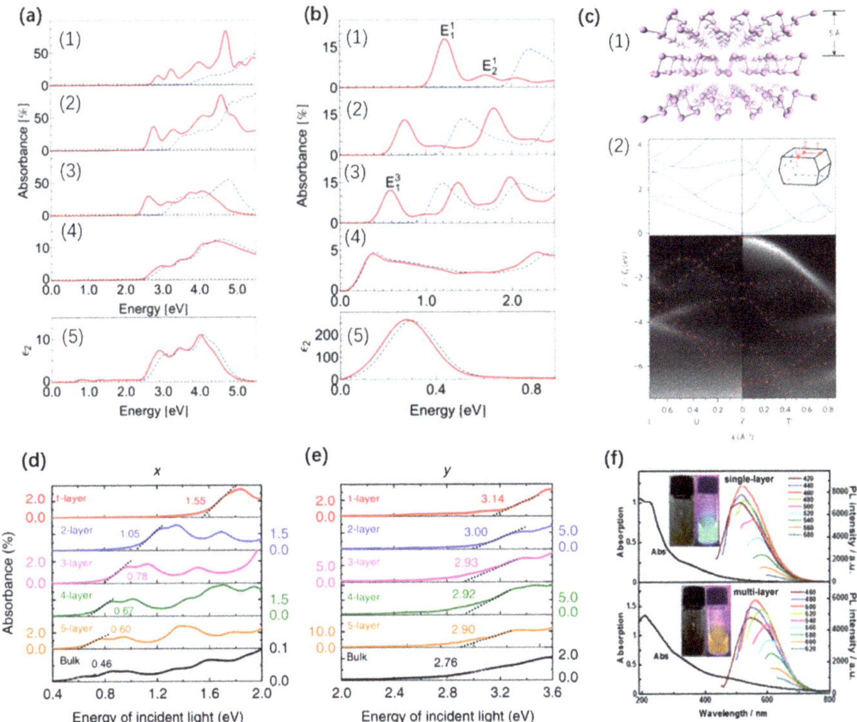

Figure 9. (**a**) Absorption spectra of monolayer (1), bilayer (2), trilayer (3), and bulk phosphorene [(4) and (5)] for the incident light polarized along the x (armchair) direction. Reprinted with permission from Ref. [135]. Copyright 2014 American Physical Society. (**b**) Optical absorption spectra of monolayer (1), bilayer (2), trilayer (3), and bulk phosphorene [(4) and (5)] for the incident light polarized along the x (armchair) direction. Reprinted with permission from Ref. [135]. Copyright 2014 American Physical Society. (**c**) Crystal and electronic structure of bulk black phosphorus. (1) Atomic structure of black phosphorus. (2) Band structure of bulk black phosphorus. Reprinted with permission from Ref. [134]. Copyright 2014 Springer Nature. (**d**,**e**) Optical absorption spectra of few-layer BP for light incidents in the c (z) direction and polarized along the a (x) and b (y) directions. Reprinted with permission from Ref. [136]. Copyright 2014 Springer Nature. (**f**) UV-vis absorption and PL emission spectra of GQDs1 and GQDs2 in water solution. Reprinted with permission from Ref. [131]. Copyright 2012 ROYAL SOCIETY OF CHEMISTRY.

For the preparation of mono or by-layer graphene, one of the most common methods is mechanical exfoliation of small mesas of HOPG. Firstly, one side of the HOPG sample is cleaved using scotch tape technology to obtain a clean surface. Then, the other side of the sample is glued to the copper electrode with silver epoxy. As shown in Figure 10b, this copper electrode is connected to the positive terminal of the power supply. The other copper electrode is connected to the ground terminal of the power supply. The second copper electrode is placed with a mica sheet (0.1 mm thick) and a substrate (from top to bottom, a 300 nm-thick SiO_2 layer, s 500 µm-thick silicon substrate and a 1 mm glass microscope slide) in turn. When given different voltages, graphene sheets of corresponding thicknesses are cleaved onto the SiO_2 layer from the HOPG [138].

Regarding the insulator surface of the epitaxial growth of graphene, SiC is a common choice. High-temperature sublimation of few atomic layers of Si from a mono crystalline SiC substrate is considered the best way to fabricate Few Layer Graphene (FLG) on a full wafer for industrial purposes at present [139]. Epitaxial growth on a substrate is highly dependent on the surface quality of the substrate, because very small defects can affect the growth of the film and reduce its quality. However, commercial SiC substrates have a large number of deep scratches caused by polishing damage. As a result, in order to obtain a good epitaxial growth layer, pretreatment of substrates to eliminate these scratches is particularly important. Hydrogen etching proves to be a good way to eliminate scratches [140]. After H_2 etching, samples are heated by electron bombardment in an ultra-high vacuum in order to remove the oxide. After verifying that the oxide is removed, samples are heated to temperatures ranging from 1250 °C to 1450 °C for 1–20 min. Under these conditions, thin graphite layers are formed, with the layer thickness determined predominantly by the temperature [141].

The third method has many similarities with the second one, and the major difference lies in the choice of substrate. According to Fogarassy's idea, graphene is deposited by chemical vapor deposition (CVD) on a Ni (111) thin layer substrate, which is prepared by sputtering Ni on a single crystal sapphire (0001) in an ultra-high vacuum (UHV) [142]. Since a sapphire's surface is difficult to completely clean, grains may be found with 30° rotation in the nickel thin layers. The nickel substrate is then annealed in a hydrogen atmosphere. Finally, the CVD experiments are performed in an atmosphere of a mixture of methane, argon and hydrogen [143].

In 2009, Subrahmanyam et al. introduced a new method to prepare graphene, which can produce 2–4 layers in a relatively large area. Subrahmanyam et al. made a variety of attempts with gas proportions, electrode sizes, discharge currents, etc., in the arc chamber. Only one typical experimental design is introduced here. A graphite rod (Alfa Aesar with 99.999% purity, 6 mm in diameter and 50 mm long) used as the anode and another graphite rod (13 mm in diameter and 60 mm in length) used as the cathode are both placed in a water-cooled stainless-steel chamber filled with a mixture of hydrogen and helium, and the direct current arc discharge of graphite evaporation is to be carried out in this chamber [144]. The discharge current is within the 100–150 A range, with a maximum open circuit voltage of 60 V [145]. The arc is maintained by continuously shifting the cathode at a constant distance from the anode. At the end of the experiment, the deposits collected on the walls of the arc chamber contain only graphene flakes, whereas the cathode deposits contain other impurities [144].

5.2.2. Black Phosphorus Preparation

Many methods of preparation of black phosphorus have been reported so far. The synthesis method of black phosphorus can be traced back to 1914. Bridgman et al. placed white phosphorus in a high-pressure cylinder under kerosene. The cylinder pressure was increased from 0.6 GPa to 1.2 GPa, and the temperature was raised from room temperature to 200 °C. After waiting 5 to 30 min, cooling the cylinder and releasing the pressure, the converted black phosphorus could be obtained [146].

Furthermore, Maruyama et al. prepared black phosphorus single crystal in liquid bismuth using white phosphorus as a raw material. The melted bismuth was poured over the white phosphorus and shaken, and then it was maintained at 400 °C for 20 h and was finally slowly brought to room temperature. The solid bismuth was dissolved with nitric acid, and acicular single crystal black phosphorus was obtained in the remaining solution [147].

Shirotani et al. carried out experiments with a wedge-type cubic anvil high-pressure apparatus developed by Wakatuki. They successfully prepared a large black phosphorus single crystal from red phosphorus under a high temperature and high pressure [148].

High pressure was considered a necessary condition for the preparation of black phosphorus, until recently, when reliable references of low-pressure preparation of black phosphorus were reported. Lange et al. reported a low-pressure method for preparing black phosphorus. They took red phosphorus as a raw material and added a small amount of Au, Sn and a catalytic amount of SnI_4 [149]. Nilges et al. gave a more detailed explanation of the method in his article [150]. Köpf et al. further simplified the experiment based on red phosphorus and described the detailed experimental steps in his article [151]. Refer to Figure 10 for the schematic diagram for the preparation of graphene and black phosphorus.

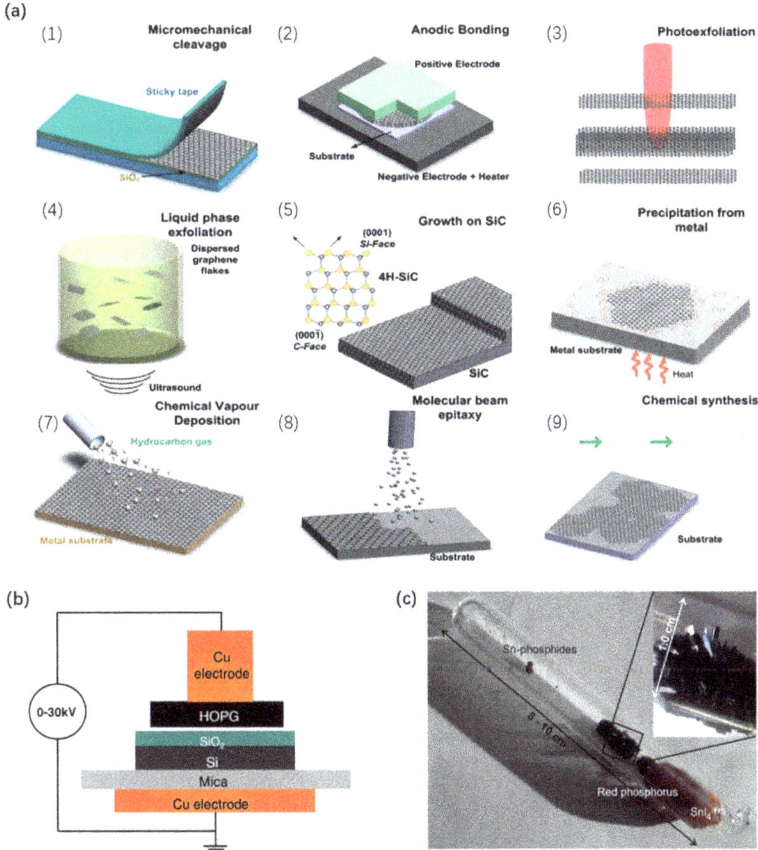

Figure 10. (**a**) Schematic illustration of the main graphene production techniques. (1) Micromechanical cleavage. (2) Anodic bonding. (3) Photoexfoliation. (4) Liquid phase exfoliation. (5) Growth on SiC. Gold and grey spheres represent Si and C atoms, respectively. At elevated T, Si atoms evaporate (arrows), leaving a carbon-rich surface that forms graphene sheets. (6) Segregation/precipitation from carbon-containing metal substrate. (7) Chemical vapor deposition. (8) Molecular beam epitaxy. (9) Chemical synthesis using benzene as a building block. Reprinted with permission from Ref. [152]. Copyright 2012 Elsevier. (**b**) Schematic of the experimental setup for electrostatic deposition of graphene sheets. Reprinted with permission from Ref. [138]. Copyright 2007 IOP Publishing. (**c**) A representative silica glass ampoule after the synthesis of black phosphorus. Reprinted with permission from Ref. [151]. Copyright 2014 Elsevier.

5.3. Devices

As the hottest material in recent years, graphene has quite a number of applications, such as touchscreen displays [153] (Figure 11f), flexible electronic devices [154] (Figure 11c), organic light-emitting diodes (OLEDs) [155] (Figure 11a,b), high-frequency transistors [156], optical modulators [157], solar cells [158] and ultra-wideband photodetectors [159].

Because monolayer graphene is a zero-band gap material, it can generate photocarriers across a wide electromagnetic spectrum. In the past few years, various graphene-based photodetectors have been reported, such as metal–graphene–metal (MGM) photodetectors, graphene p–n junction photodetectors, graphene–semiconductor heterojunction photodetectors and hybrid photodetectors [159].

As another representative 2D material, black phosphorus has great potential in FET, photoelectric detectors, battery anode materials and thermoelectric applications due to its unique structure and rare anisotropy [160]. In 2014, Liu et al. made the first report on FET based on black-phosphorus-layered materials [161]. It has become widespread research due to its ambipolar FET [162]. Drain-current modulation at room temperature is four orders of magnitude higher than graphene [163] (Figure 11e,g). In addition, black phosphorus has the potential to be used in flexible and wearable products due to its binary properties [164]. Zhu et al. reported a flexible bipolar transistor circuit and an AM demodulator based on black phosphorus [165] (Figure 11d). The low field hole mobility was 310 $cm^2\ V^{-1}s^{-1}$, which is much higher than the most advanced metal oxide flexible transistors.

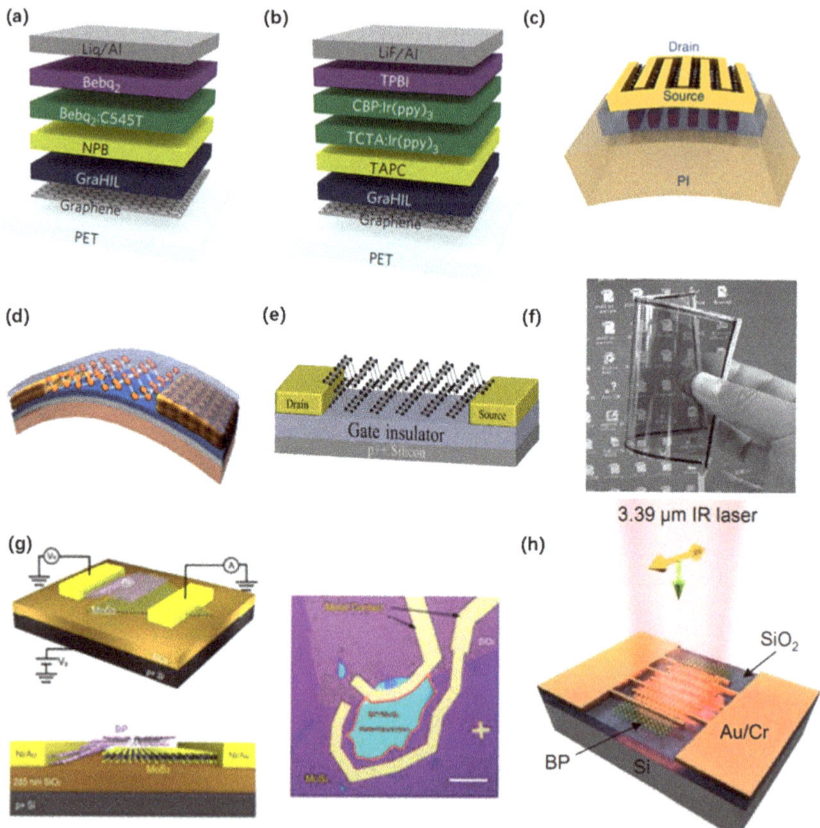

Figure 11. (**a**,**b**) Device structures of flexible small-molecule fluorescent OLEDs (**a**) and phosphorescent OLEDs (**b**) using a graphene anode modified with a GraHIL. Reprinted with permission from Ref. [155]. Copyright 2012 Springer Nature. (**c**) Illustration of multifinger embedded-gate graphene device structure that can afford high current drive, thin gate dielectrics, low gate resistance, surface passivation and simple post-transfer fabrication. Reprinted with permission from Ref. [154]. Copyright 2014 Springer Nature. (**d**) Simplified illustration of the device structure of flexible BG BP FET on PI substrate (not to scale). Reprinted with permission from Ref. [165]. Copyright 2015 American

Chemical Society. (**e**) Schematic diagram of the back-gated BP FET. Reprinted with permission from Ref. [163]. Copyright 2015 Royal Society of Chemistry. (**f**) A flexible four-wire resistance touch screen fabricated using RGO/PET as electrodes. Reprinted with permission from Ref. [153]. Copyright 2012 John Wiley and Sons. (**g**) Schematics and optical image of the device structure. Reprinted with permission from Ref. [163]. Copyright 2015 Royal Society of Chemistry. (**h**) Schematic diagram of the BP MSM photodetector operating at 3.39 μm. Reprinted with permission from Ref. [16]. Copyright 2016 American Chemical Society.

It is worth mentioning that mixed-dimensional materials can be a solution in this field. A photodetector based on graphene/$PdSe_2$/germanium heterojunctions has been recently reported. Mixed-dimensional heterojunctions provide enhanced light absorption, and graphene electrodes provide effective carrier collection. Owing to above advantages, this photodetector exhibits great device performance, such as a high specific detectivity and broadband photosensitivity from ultraviolet to MWIR.

In order to better reflect the properties of photodetectors based on different materials, Table 1 shows the performance comparison between the most representative devices in the above mentioned.

Table 1. Performance comparison of several representative devices.

Device	R(A/W)	D*(Jones)	Response Time	Reference
Graphene–PbS QD phototransistors	~10^8	7×10^3	NA	[46]
PbS CQD–TMDC photodetectors	10^7	1.02×10^{12}	NA	[50]
Tandem PbSe CQD photodetectors	~0.5	8.1×10^{13}	NA	[83]
PbSe CQD–Bi2O2Se nanosheets hybrid photodetector	>10^3	NA	~4 ms	[87]
HgTe CQD photodetectors	1.3	3.3×10^{11}	50 ns	[117]
Stacked CQD photodiodes	NA	6×10^{10}	<2.5 μs	[11]
Three-pixel photodetectors	0.1	2×10^7	~ 91.5 ms and ~541.3 ms	[108]
HgTe CQD/Graphene/Silicon devices	~0.9	~5×10^9	~13 ns and ~3 μs	[120]

6. Conclusions

In this paper, the development of room-temperature infrared detectors based on low-dimensional materials is reviewed. We mainly introduce the development of PbS CQDs, PbSe CQDs, HgTe CQDs and some 2D materials. The properties, preparations and device developments of each material are described in detail.

Room-temperature infrared photodetectors theoretically solve a series of problems, such as large volume, high cost and inconvenient use of refrigeration detectors. In addition, the liquid phase preparation of low-dimensional materials also allows low-dimensional materials to be integrated with traditional silicon electronics and flexible large-area substrates at a low cost and with high safety. Additionally, there are still many challenges for IR photodetectors in this field. For example, it is still difficult to achieve a consistent control effect on material synthesis, so the development of precursor chemistry is very important. In addition, learning how to improve multiple-exciton generation (MEG)-enhanced quantum efficiency is also a challenge of great concern. Although there are still many problems to be solved for material growth modes and circuit coupling modes, low-dimensional materials still have broad application prospects, such as subwavelength pixels, large arrays and multicolor devices. Room-temperature infrared photodetectors with low-dimensional materials have already shown great potential in commercialization.

Author Contributions: Conceptualization, X.T. and M.C.; investigation, T.L.; writing—original draft preparation, T.L.; writing—review and editing, M.C. and X.T.; project administration, M.C.; funding acquisition, X.T. All authors have read and agreed to the published version of the manuscript.

Funding: This research was funded by the National Natural Science Foundation of China and the National Key R&D Program of China (NSFC No. 62035004, 2021YFA0717600 and NSFC No.62105022).

Institutional Review Board Statement: Not applicable.

Informed Consent Statement: Not applicable.

Data Availability Statement: Not applicable.

Conflicts of Interest: The authors declare no conflict of interest.

References

1. Rogalski, A. *Infrared Detectors*; CRC Press: Boca Raton, FL, USA, 2000.
2. Rogalski, A.; Martyniuk, P.; Kopytko, M. Challenges of small-pixel infrared detectors: A review. *IOP Sci.* **2016**, *79*, 046501. [CrossRef]
3. Konstantatos, G.; Sargent, E.H. Nanostructured materials for photon detection. *Nat. Nanotechnol.* **2010**, *5*, 391–400. [CrossRef]
4. Wang, J.; Hu, W. Recent progress on integrating two-dimensional materials with ferroelectrics for memory devices and photodetectors. *Chin. Phys. B* **2017**, *26*, 037106. [CrossRef]
5. Zhang, S.; Hu, Y.; Hao, Q. Advances of sensitive infrared detectors with HgTe colloidal quantum dots. *Coatings* **2020**, *10*, 760. [CrossRef]
6. Kagan, C.R.; Lifshitz, E.; Sargent, E.H.; Talapin, D.V. Building devices from colloidal quantum dots. *Science* **2016**, *353*, aac5523. [CrossRef]
7. Litvin, A.; Martynenko, I.; Purcell-Milton, F.; Baranov, A.; Fedorov, A.; Gun'Ko, Y.J. Colloidal quantum dots for optoelectronics. *J. Mater. Chem. A* **2017**, *5*, 13252–13275. [CrossRef]
8. Lhuillier, E.; Scarafagio, M.; Hease, P.; Nadal, B.; Aubin, H.; Xu, X.Z.; Lequeux, N.; Patriarche, G.; Ithurria, S.; Dubertret, B. Infrared photodetection based on colloidal quantum-dot films with high mobility and optical absorption up to THz. *Nano Lett.* **2016**, *16*, 1282–1286. [CrossRef]
9. Pietryga, J.M.; Park, Y.-S.; Lim, J.; Fidler, A.F.; Bae, W.K.; Brovelli, S.; Klimov, V.I. Spectroscopic and device aspects of nanocrystal quantum dots. *Chem. Rev.* **2016**, *116*, 10513–10622. [CrossRef]
10. Wang, R.; Shang, Y.; Kanjanaboos, P.; Zhou, W.; Ning, Z.; Sargent, E.H.E. Colloidal quantum dot ligand engineering for high performance solar cells. *Energy Environ. Sci.* **2016**, *9*, 1130–1143. [CrossRef]
11. Tang, X.; Ackerman, M.M.; Chen, M.; Guyot-Sionnest, P. Dual-band infrared imaging using stacked colloidal quantum dot photodiodes. *Nat. Photonics* **2019**, *13*, 277–282. [CrossRef]
12. Tang, X.; Chen, M.; Ackerman, M.M.; Melnychuk, C.; Guyot-Sionnest, P. Direct Imprinting of Quasi-3D Nanophotonic Structures into Colloidal Quantum-Dot Devices. *Adv. Mater.* **2020**, *32*, 1906590. [CrossRef] [PubMed]
13. Shen, G.; Chen, D. One-dimensional nanostructures for photodetectors. *Recent Pat. Nanotechnol.* **2010**, *4*, 20–31. [CrossRef] [PubMed]
14. Miao, J.; Hu, W.; Guo, N.; Lu, Z.; Zou, X.; Liao, L.; Shi, S.; Chen, P.; Fan, Z.; Ho, J.C. Single InAs nanowire room-temperature near-infrared photodetectors. *ACS Nano* **2014**, *8*, 3628–3635. [CrossRef] [PubMed]
15. Gan, X.; Shiue, R.-J.; Gao, Y.; Meric, I.; Heinz, T.F.; Shepard, K.; Hone, J.; Assefa, S.; Englund, D. Chip-integrated ultrafast graphene photodetector with high responsivity. *Nat. Photonics* **2013**, *7*, 883–887. [CrossRef]
16. Guo, Q.; Pospischil, A.; Bhuiyan, M.; Jiang, H.; Tian, H.; Farmer, D.; Deng, B.; Li, C.; Han, S.-J.; Wang, H. Black phosphorus mid-infrared photodetectors with high gain. *Nano Lett.* **2016**, *16*, 4648–4655. [CrossRef]
17. Mannhart, J.; Blank, D.H.; Hwang, H.; Millis, A.; Triscone, J.-M. Two-dimensional electron gases at oxide interfaces. *MRS Bull.* **2008**, *33*, 1027–1034. [CrossRef]
18. Glazov, M.; Sherman, E.Y.; Dugaev, V. Two-dimensional electron gas with spin–orbit coupling disorder. *Phys. E Low-Dimens. Syst. Nanostruct.* **2010**, *42*, 2157–2177. [CrossRef]
19. Bristowe, N.; Ghosez, P.; Littlewood, P.B.; Artacho, E.J. The origin of two-dimensional electron gases at oxide interfaces: Insights from theory. *J. Phys. Condens. Matter* **2014**, *26*, 143201. [CrossRef]
20. Gupta, M.C.; Harrison, J.T.; Islam, M.T. Photoconductive PbSe thin films for infrared imaging. *Mater. Adv.* **2021**, *2*, 3133–3160. [CrossRef]
21. Wang, P.; Xia, H.; Li, Q.; Wang, F.; Zhang, L.; Li, T.; Martyniuk, P.; Rogalski, A.; Hu, W. Sensing infrared photons at room temperature: From bulk materials to atomic layers. *Nano Micro Small* **2019**, *15*, 1904396. [CrossRef]
22. Liu, T.; Tong, L.; Huang, X.; Ye, L.J. Room-temperature infrared photodetectors with hybrid structure based on two-dimensional materials. *Chin. Phys. B* **2019**, *28*, 017302. [CrossRef]
23. Tan, C.L.; Mohseni, H. Emerging technologies for high performance infrared detectors. *Nanophotonics* **2018**, *7*, 169–197. [CrossRef]
24. Case, T.W. Notes on the change of resistance of certain substances in light. *Phys. Rev.* **1917**, *9*, 305. [CrossRef]
25. Case, T.W. "Thalofide Cell"—A New Photo-Electric Substance. *Phys. Rev.* **1920**, *15*, 289. [CrossRef]
26. Johnson, T. Lead salt detectors and arrays PbS and PbSe. In Proceedings of the Infrared Detectors, San Diego, CA, USA, 9 December 1983; pp. 60–94.
27. Scholes, G.D.; Rumbles, G. Excitons in nanoscale systems. *Mater. Sustain. Energy* **2011**, 12–25.
28. Sadovnikov, S.I.; Gusev, A.I.; Rempel, A.A. Nanostructured lead sulfide: Synthesis, structure and properties. *Russ. Chem. Rev.* **2016**, *85*, 731. [CrossRef]

29. Yang, Y.J.; He, L.Y.; Zhang, Q.F. A cyclic voltammetric synthesis of PbS nanoparticles. *Electrochem. Commun.* **2005**, *7*, 361–364. [CrossRef]
30. Ukhanov, I.I. Optical properties of semiconductors. *Moscow Izdatel Nauka.* **1977**, 368.
31. Elliott, R. Intensity of optical absorption by excitons. *Phys. Rev.* **1957**, *108*, 1384. [CrossRef]
32. Sadovnikov, S.; Kozhevnikova, N.; Gusev, A. Optical properties of nanostructured lead sulfide films with a D03 cubic structure. *Semiconductors* **2011**, *45*, 1559–1570. [CrossRef]
33. Peterson, J.J.; Krauss, T.D. Fluorescence spectroscopy of single lead sulfide quantum dots. *Nano Lett.* **2006**, *6*, 510–514. [CrossRef] [PubMed]
34. Hines, M.A.; Scholes, G.D. Colloidal PbS nanocrystals with size-tunable near-infrared emission: Observation of post-synthesis self-narrowing of the particle size distribution. *Adv. Mater.* **2003**, *15*, 1844–1849. [CrossRef]
35. Brown, P.R.; Kim, D.; Lunt, R.R.; Zhao, N.; Bawendi, M.G.; Grossman, J.C.; Bulovic, V. Energy level modification in lead sulfide quantum dot thin films through ligand exchange. *ACS Nano* **2014**, *8*, 5863–5872. [CrossRef]
36. Tang, H.; Zhong, J.; Chen, W.; Shi, K.; Mei, G.; Zhang, Y.; Wen, Z.; Muller-Buschbaum, P.; Wu, D.; Wang, K. Lead sulfide quantum dot photodetector with enhanced responsivity through a two-step ligand-exchange method. *ACS Appl. Nano Mater.* **2019**, *2*, 6135–6143. [CrossRef]
37. Ushakova, E.V.; Cherevkov, S.A.; Litvin, A.P.; Parfenov, P.S.; Volgina, D.-O.A.; Kasatkin, I.A.; Fedorov, A.V.; Baranov, A.V. Ligand-dependent morphology and optical properties of lead sulfide quantum dot superlattices. *J. Phys. Chem. C* **2016**, *120*, 25061–25067. [CrossRef]
38. Sashchiuk, A.; Lifshitz, E.; Reisfeld, R.; Saraidarov, T.; Zelner, M.; Willenz, A.J. Optical and conductivity properties of PbS nanocrystals in amorphous zirconia sol-gel films. *J. Sol-Gel Sci. Technol.* **2002**, *24*, 31–38. [CrossRef]
39. Tang, J.; Brzozowski, L.; Barkhouse, D.A.R.; Wang, X.; Debnath, R.; Wolowiec, R.; Palmiano, E.; Levina, L.; Pattantyus-Abraham, A.G.; Jamakosmanovic, D. Quantum dot photovoltaics in the extreme quantum confinement regime: The surface-chemical origins of exceptional air-and light-stability. *ACS Nano* **2010**, *4*, 869–878. [CrossRef]
40. Cademartiri, L.; von Freymann, G.; Arsenault, A.C.; Bertolotti, J.; Wiersma, D.S.; Kitaev, V.; Ozin, G.A. Nanocrystals as precursors for flexible functional films. *Small* **2005**, *1*, 1184–1187. [CrossRef]
41. McDonald, S.A.; Konstantatos, G.; Zhang, S.; Cyr, P.W.; Klem, E.J.; Levina, L.; Sargent, E.H. Solution-processed PbS quantum dot infrared photodetectors and photovoltaics. In *Materials for Sustainable Energy: A Collection of Peer-Reviewed Research and Review Articles from Nature Publishing Group*; World Scientific: Singapore, 2011; pp. 70–74.
42. Onishchuk, D.; Pavlyuk, A.; Parfenov, P.; Litvin, A.; Nabiev, I. Near Infrared LED Based on PbS Nanocrystals. *Opt. Spectrosc.* **2018**, *125*, 751–755. [CrossRef]
43. Ramiro, I.I.; Ozdemir, O.; Christodoulou, S.; Gupta, S.; Dalmases, M.; Torre, I.; Konstantatos, G. Mid-and long-wave infrared optoelectronics via intraband transitions in PbS colloidal quantum dots. *Nano Lett.* **2020**, *20*, 1003–1008. [CrossRef]
44. Zhou, W.; Shang, Y.; García de Arquer, F.P.; Xu, K.; Wang, R.; Luo, S.; Xiao, X.; Zhou, X.; Huang, R.; Sargent, E.H. Solution-processed upconversion photodetectors based on quantum dots. *Nat. Electron.* **2020**, *3*, 251–258. [CrossRef]
45. Balazs, D.M.; Rizkia, N.; Fang, H.-H.; Dirin, D.N.; Momand, J.; Kooi, B.J.; Kovalenko, M.V.; Loi, M.A. Interfaces. Colloidal quantum dot inks for single-step-fabricated field-effect transistors: The importance of postdeposition ligand removal. *ACS Appl. Mater. Interfaces* **2018**, *10*, 5626–5632. [CrossRef] [PubMed]
46. Konstantatos, G.; Badioli, M.; Gaudreau, L.; Osmond, J.; Bernechea, M.; De Arquer, F.; Gatti, F.; Koppens, F.H. Hybrid graphene–quantum dot phototransistors with ultrahigh gain. *Nat. Nanotechnol.* **2012**, *7*, 363–368. [CrossRef]
47. Sun, Z.; Liu, Z.; Li, J.; Tai, G.A.; Lau, S.P.; Yan, F. Infrared photodetectors based on CVD-grown graphene and PbS quantum dots with ultrahigh responsivity. *Adv. Mater.* **2012**, *24*, 5878–5883. [CrossRef] [PubMed]
48. Che, Y.; Zhang, Y.; Cao, X.; Song, X.; Zhang, H.; Cao, M.; Dai, H.; Yang, J.; Zhang, G.; Yao, J. High-performance PbS quantum dot vertical field-effect phototransistor using graphene as a transparent electrode. *Appl. Phys. Lett.* **2016**, *109*, 263101. [CrossRef]
49. Hwang, D.K.; Lee, Y.T.; Lee, H.S.; Lee, Y.J.; Shokouh, S.H.; Kyhm, J.-H.; Lee, J.; Kim, H.H.; Yoo, T.-H.; Nam, S.H. Ultrasensitive PbS quantum-dot-sensitized InGaZnO hybrid photoinverter for near-infrared detection and imaging with high photogain. *NPG Asia Mater.* **2016**, *8*, e233. [CrossRef]
50. Ozdemir, O.; Ramiro, I.; Gupta, S.; Konstantatos, G.J. High sensitivity hybrid PbS CQD-TMDC photodetectors up to 2 μm. *Nat. Photonics* **2019**, *6*, 2381–2386. [CrossRef]
51. YousefiAmin, A.; Killilea, N.A.; Sytnyk, M.; Maisch, P.; Tam, K.C.; Egelhaaf, H.-J.; Langner, S.; Stubhan, T.; Brabec, C.J.; Rejek, T.J. Fully printed infrared photodetectors from PbS nanocrystals with perovskite ligands. *ACS Nano* **2019**, *13*, 2389–2397. [CrossRef]
52. Peng, M.; Wang, Y.; Shen, Q.; Xie, X.; Zheng, H.; Ma, W.; Wen, Z.; Sun, X.J.S.C.M. High-performance flexible and broadband photodetectors based on PbS quantum dots/ZnO nanoparticles heterostructure. *Sci. China Mater.* **2019**, *62*, 225–235. [CrossRef]
53. Ren, Y.; Dai, T.; Luo, W.; Liu, X.J. Evidences of sensitization mechanism for PbSe thin films photoconductor. *Vacuum* **2018**, *149*, 190–194. [CrossRef]
54. Qiu, J.; Weng, B.; Yuan, Z.; Shi, Z.J. Study of sensitization process on mid-infrared uncooled PbSe photoconductive detectors leads to high detectivity. *J. Appl. Phys.* **2013**, *113*, 103102. [CrossRef]
55. Shandalov, M.; Golan, Y.J.T. Microstructure and morphology evolution in chemical solution deposited semiconductor films: 2. PbSe on as face of GaAs (111). *Eur. Phys. J. Appl. Phys.* **2004**, *28*, 51–57. [CrossRef]

56. Wise, F.W. Lead salt quantum dots: The limit of strong quantum confinement. *Acc. Chem. Res.* **2000**, *33*, 773–780. [CrossRef] [PubMed]
57. Begum, A.; Hussain, A.; Rahman, A.J. Effect of deposition temperature on the structural and optical properties of chemically prepared nanocrystalline lead selenide thin films. *Beilstein J. Nanotechnol.* **2012**, *3*, 438–443. [CrossRef]
58. Joshi, R.K.; Kanjilal, A.; Sehgal, H. Size dependence of optical properties in solution-grown Pb1−xFexS nanoparticle films. *Nanotechnology* **2003**, *14*, 809. [CrossRef]
59. Zhu, J.; Aruna, S.; Koltypin, Y.; Gedanken, A. A novel method for the preparation of lead selenide: Pulse sonoelectrochemical synthesis of lead selenide nanoparticles. *Chem. Mater.* **2000**, *12*, 143–147. [CrossRef]
60. Gao, J.; Fidler, A.F.; Klimov, V.I. Carrier multiplication detected through transient photocurrent in device-grade films of lead selenide quantum dots. *Nat. Commun.* **2015**, *6*, 8185. [CrossRef]
61. Thambidurai, M.; Jang, Y.; Shapiro, A.; Yuan, G.; Xiaonan, H.; Xuechao, Y.; Wang, Q.J.; Lifshitz, E.; Demir, H.V.; Dang, C. High performance infrared photodetectors up to 2.8 μm wavelength based on lead selenide colloidal quantum dots. *Opt. Mater. Express* **2017**, *7*, 2326–2335. [CrossRef]
62. Ahmad, W.; He, J.; Liu, Z.; Xu, K.; Chen, Z.; Yang, X.; Li, D.; Xia, Y.; Zhang, J.; Chen, C. Lead selenide (PbSe) colloidal quantum dot solar cells with >10% efficiency. *Adv. Mater.* **2019**, *31*, 1900593. [CrossRef]
63. Hodes, G. *Chemical Solution Deposition of Semiconductor Films*; CRC Press: Boca Raton, FL, USA, 2002.
64. Jang, M.-H.; Hoglund, E.R.; Litwin, P.M.; Yoo, S.-S.; McDonnell, S.J.; Howe, J.M.; Gupta, M.C. Photoconductive mechanism of IR-sensitive iodized PbSe thin films via strong hole–phonon interaction and minority carrier diffusion. *Appl. Opt.* **2020**, *59*, 10228–10235. [CrossRef]
65. Jang, M.-H.; Yoo, S.-S.; Kramer, M.T.; Dhar, N.K.; Gupta, M.C. Properties of chemical bath deposited and sensitized PbSe thin films for IR detection. *Semicond. Sci. Technol.* **2019**, *34*, 115010. [CrossRef]
66. Hankare, P.; Delekar, S.; Bhuse, V.; Garadkar, K.; Sabane, S.; Gavali, L. Synthesis and characterization of chemically deposited lead selenide thin films. *Mater. Chem. Phys.* **2003**, *82*, 505–508. [CrossRef]
67. Kassim, A.; Min, H.S.; Nagalingam, S.J. Preparation of Lead Selenide Thin Films by Chemical Bath Deposition Method in The Presence of Complexing Agent (Tartaric Acid). *Turk. J. Sci. Technol.* **2011**, *6*, 17–23.
68. Anuar, K.; Abdul, H.; Ho, S.; Saravanan, N. Effect of deposition time on surface topography of chemical bath deposited PbSe thin films observed by atomic force microscopy. *Pac. J. Sci. Technol.* **2010**, *11*, 399–403.
69. Hone, F.G.; Ampong, F.K.; Abza, T.; Nkrumah, I.; Paal, M.; Nkum, R.K.; Boakye, F. The effect of deposition time on the structural, morphological and optical band gap of lead selenide thin films synthesized by chemical bath deposition method. *Mater. Lett.* **2015**, *155*, 58–61. [CrossRef]
70. Hone, F.G.; Ampong, F.K. Effect of deposition temperature on the structural, morphological and optical band gap of lead selenide thin films synthesized by chemical bath deposition method. *Mater. Chem. Phys.* **2016**, *183*, 320–325. [CrossRef]
71. Ghobadi, N.; Hatam, E.G.J. Surface studies, structural characterization and quantity determination of PbSe nanocrystals deposited by chemical bath deposition technique. *J. Cryst. Growth* **2015**, *418*, 111–114. [CrossRef]
72. Harrison, J.T.; Pantoja, E.; Jang, M.-H.; Gupta, M.C.J. Laser sintered PbSe semiconductor thin films for Mid-IR applications using nanocrystals. *J. Alloy. Compd.* **2020**, *849*, 156537. [CrossRef]
73. Jadhav, S.; Khairnar, U. Study of optical properties of co-evaporated PbSe thin films. *Arch. Appl. Sci. Res.* **2012**, *4*, 169–177.
74. Shyju, T.; Anandhi, S.; Sivakumar, R.; Garg, S.; Gopalakrishnan, R.J. Investigation on structural, optical, morphological and electrical properties of thermally deposited lead selenide (PbSe) nanocrystalline thin films. *J. Cryst. Growth* **2012**, *353*, 47–54. [CrossRef]
75. Rumianowski, R.T.; Dygdala, R.S.; Jung, W.; Bala, W.J. Growth of PbSe thin films on Si substrates by pulsed laser deposition method. *J. Cryst. Growth* **2003**, *252*, 230–235. [CrossRef]
76. Thanh, N.T.; Maclean, N.; Mahiddine, S. Mechanisms of nucleation and growth of nanoparticles in solution. *Chem. Rev.* **2014**, *114*, 7610–7630. [CrossRef] [PubMed]
77. Kumar, S.; Nann, T.J. Shape control of II–VI semiconductor nanomaterials. *Small* **2006**, *2*, 316–329. [CrossRef] [PubMed]
78. Burda, C.; Chen, X.; Narayanan, R.; El-Sayed, M.A. Chemistry and properties of nanocrystals of different shapes. *Chem. Rev.* **2005**, *105*, 1025–1102. [CrossRef] [PubMed]
79. Murray, C.B.; Sun, S.; Gaschler, W.; Doyle, H.; Betley, T.A.; Kagan, C.R. Colloidal synthesis of nanocrystals and nanocrystal superlattices. *IBM J. Res. Dev.* **2001**, *45*, 47–56. [CrossRef]
80. Huang, Z.; Gao, M.; Yan, Z.; Pan, T.; Liao, F.; Lin, Y. Flexible infrared detectors based on p–n junctions of multi-walled carbon nanotubes. *Nanoscale* **2016**, *8*, 9592–9599. [CrossRef]
81. Nakotte, T.; Luo, H.; Pietryga, J. PbE (E = S, Se) colloidal quantum dot-layered 2D material hybrid photodetectors. *Nanomaterials* **2020**, *10*, 172. [CrossRef]
82. Borousan, F.; Shabani, P.; Yousefi, R. Improvement of visible-near-infrared (NIR) broad spectral photocurrent application of PbSe mesostructures using tuning the morphology and optical properties. *Mater. Res. Express* **2019**, *6*, 095016. [CrossRef]
83. Jiang, Z.; Hu, W.; Mo, C.; Liu, Y.; Zhang, W.; You, G.; Wang, L.; Atalla, M.R.; Zhang, Y.; Liu, J. Ultra-sensitive tandem colloidal quantum-dot photodetectors. *Nanoscale* **2015**, *7*, 16195–16199. [CrossRef]

84. Sulaman, M.; Song, Y.; Yang, S.; Hao, Q.; Zhao, Y.; Li, M.; Saleem, M.I.; Chandraseakar, P.V.; Jiang, Y.; Tang, Y. High-performance solution-processed colloidal quantum dots-based tandem broadband photodetectors with dielectric interlayer. *Nanotechnology* **2019**, *30*, 465203. [CrossRef]
85. Che, Y.; Cao, X.; Yao, J. A PbSe nanocrystal vertical phototransistor with graphene electrode. *Opt. Mater.* **2019**, *89*, 138–141. [CrossRef]
86. Zhang, Y.; Cao, M.; Song, X.; Wang, J.; Che, Y.; Dai, H.; Ding, X.; Zhang, G.; Yao, J.J.T. Multiheterojunction phototransistors based on graphene–PbSe quantum dot hybrids. *J. Phys. Chem. C* **2015**, *119*, 21739–21743. [CrossRef]
87. Luo, P.; Zhuge, F.; Wang, F.; Lian, L.; Liu, K.; Zhang, J.; Zhai, T. PbSe quantum dots sensitized high-mobility Bi2O2Se nanosheets for high-performance and broadband photodetection beyond 2 μm. *ACS Nano* **2019**, *13*, 9028–9037. [CrossRef] [PubMed]
88. Keuleyan, S.; Lhuillier, E.; Guyot-Sionnest, P.J. Synthesis of colloidal HgTe quantum dots for narrow mid-IR emission and detection. *J. Am. Chem. Soc.* **2011**, *133*, 16422–16424. [CrossRef]
89. Keuleyan, S.; Lhuillier, E.; Brajuskovic, V.; Guyot-Sionnest, P. Mid-infrared HgTe colloidal quantum dot photodetectors. *Nat. Photonics* **2011**, *5*, 489–493. [CrossRef]
90. Keuleyan, S.E.; Guyot-Sionnest, P.; Delerue, C.; Allan, G. Mercury telluride colloidal quantum dots: Electronic structure, size-dependent spectra, and photocurrent detection up to 12 μm. *ACS Nano* **2014**, *8*, 8676–8682. [CrossRef]
91. Allan, G.; Delerue, C. Tight-binding calculations of the optical properties of HgTe nanocrystals. *Phys. Rev. B* **2012**, *86*, 165437. [CrossRef]
92. Delerue, C.J.; Lannoo, M. *Nanostructures: Theory and Modeling*; Springer Science & Business Media: Berlin, Germany, 2013.
93. Allan, G.; Delerue, C. Confinement effects in PbSe quantum wells and nanocrystals. *Phys. Rev. B* **2004**, *70*, 245321. [CrossRef]
94. Shen, G.; Chen, M.; Guyot-Sionnest, P. Synthesis of nonaggregating HgTe colloidal quantum dots and the emergence of air-stable n-doping. *J. Phys. Chem. Lett.* **2017**, *8*, 2224–2228. [CrossRef]
95. Hudson, M.H.; Chen, M.; Kamysbayev, V.; Janke, E.M.; Lan, X.; Allan, G.; Delerue, C.; Lee, B.; Guyot-Sionnest, P.; Talapin, D.V. Conduction band fine structure in colloidal HgTe quantum dots. *ACS Nano* **2018**, *12*, 9397–9404. [CrossRef]
96. Chen, M.; Lan, X.; Tang, X.; Wang, Y.; Hudson, M.H.; Talapin, D.V.; Guyot-Sionnest, P. High carrier mobility in HgTe quantum dot solids improves mid-IR photodetectors. *ACS Photonics* **2019**, *6*, 2358–2365. [CrossRef]
97. Lhuillier, E.; Keuleyan, S.; Guyot-Sionnest, P. Optical properties of HgTe colloidal quantum dots. *Nanotechnology* **2012**, *23*, 175705. [CrossRef] [PubMed]
98. Kovalenko, M.V.; Kaufmann, E.; Pachinger, D.; Roither, J.; Huber, M.; Stangl, J.; Hesser, G.; Schäffler, F.; Heiss, W.J. Colloidal HgTe nanocrystals with widely tunable narrow band gap energies: From telecommunications to molecular vibrations. *J. Am. Chem. Soc.* **2006**, *128*, 3516–3517. [CrossRef]
99. Brennan, J.; Siegrist, T.; Carroll, P.; Stuczynski, S.; Reynders, P.; Brus, L.; Steigerwald, M. Bulk and nanostructure group II-VI compounds from molecular organometallic precursors. *Chem. Mater.* **1990**, *2*, 403–409. [CrossRef]
100. Rogach, A.; Kershaw, S.V.; Burt, M.; Harrison, M.T.; Kornowski, A.; Eychmüller, A.; Weller, H. Colloidally prepared HgTe nanocrystals with strong room-temperature infrared luminescence. *Adv. Mater.* **1999**, *11*, 552–555. [CrossRef]
101. Murray, C.; Norris, D.J.; Bawendi, M.G.J. Synthesis and characterization of nearly monodisperse CdE (E= sulfur, selenium, tellurium) semiconductor nanocrystallites. *J. Am. Chem. Soc.* **1993**, *115*, 8706–8715. [CrossRef]
102. Herron, N.; Calabrese, J.; Farneth, W.; Wang, Y. Crystal structure and optical properties of Cd32S14 (SC6H5) 36. DMF4, a cluster with a 15 Angstrom CdS core. *Science* **1993**, *259*, 1426–1428. [CrossRef]
103. Vossmeyer, T.; Katsikas, L.; Giersig, M.; Popovic, I.; Diesner, K.; Chemseddine, A.; Eychmüller, A.; Weller, H.J.T. CdS nanoclusters: Synthesis, characterization, size dependent oscillator strength, temperature shift of the excitonic transition energy, and reversible absorbance shift. *J. Phys. Chem.* **1994**, *98*, 7665–7673. [CrossRef]
104. Rogach, A.; Katsikas, L.; Kornowski, A.; Su, D.; Eychmüller, A.; Weller, H.J. Synthesis and characterization of thiol-stabilized CdTe nanocrystals. *Ber. Bunsenges. Phys. Chem.* **1996**, *100*, 1772–1778. [CrossRef]
105. Huang, L.; Zingaro, R.A.; Meyers, E.A.; Reibenspies, J.H. Reaction of mercury (II) dibromide with tris (n-butyl) phosphine telluride: Formation of an unusual (Hg Te) 3 ring system. *Heteroat. Chem.* **1996**, *7*, 57–65. [CrossRef]
106. Green, M.; Wakefield, G.; Dobson, P.J. A simple metalorganic route to organically passivated mercury telluride nanocrystals. *Heteroat. Chem.* **2003**, *13*, 1076–1078. [CrossRef]
107. Cademartiri, L.; Bertolotti, J.; Sapienza, R.; Wiersma, D.S.; Von Freymann, G.; Ozin, G.A.J.T. Multigram scale, solventless, and diffusion-controlled route to highly monodisperse PbS nanocrystals. *J. Phys. Chem. B* **2006**, *110*, 671–673. [CrossRef] [PubMed]
108. Tang, X.; Tang, X.; Lai, K.W.C. Scalable fabrication of infrared detectors with multispectral photoresponse based on patterned colloidal quantum dot films. *ACS Photonics* **2016**, *3*, 2396–2404. [CrossRef]
109. Lhuillier, E.; Keuleyan, S.; Zolotavin, P.; Guyot-Sionnest, P.J. Mid-Infrared HgTe/As2S3 Field Effect Transistors and Photodetectors. *Adv. Mater.* **2013**, *25*, 137–141. [CrossRef] [PubMed]
110. Keuleyan, S.; Kohler, J.; Guyot-Sionnest, P.J.T. Photoluminescence of mid-infrared HgTe colloidal quantum dots. *J. Phys. Chem. C* **2014**, *118*, 2749–2753. [CrossRef]
111. Guyot-Sionnest, P.; Roberts, J.A. Background limited mid-infrared photodetection with photovoltaic HgTe colloidal quantum dots. *Appl. Phys. Lett.* **2015**, *107*, 253104. [CrossRef]

112. Melnychuk, C.; Guyot-Sionnest, P.J.T. Slow Auger relaxation in HgTe colloidal quantum dots. *J. Phys. Chem. Lett.* **2018**, *9*, 2208–2211. [CrossRef]
113. Tang, X.; Ackerman, M.M.; Guyot-Sionnest, P. Thermal imaging with plasmon resonance enhanced HgTe colloidal quantum dot photovoltaic devices. *ACS Nano* **2018**, *12*, 7362–7370. [CrossRef]
114. Shen, G.; Guyot-Sionnest, P. HgTe/CdTe and HgSe/CdX (X = S, Se, and Te) core/shell mid-infrared quantum dots. *Chem. Mater.* **2018**, *31*, 286–293. [CrossRef]
115. Ackerman, M.M.; Chen, M.; Guyot-Sionnest, P. HgTe colloidal quantum dot photodiodes for extended short-wave infrared detection. *Appl. Phys. Lett.* **2020**, *116*, 083502. [CrossRef]
116. Chen, M.; Lan, X.; Hudson, M.; Shen, G.; Littlewood, P.; Talapin, D.; Guyot-Sionnest, P. Magnetoresistance of High Mobility HgTe Quantum Dot Films with Controlled Charging. *J. Phys. Chem. C* **2022**. [CrossRef]
117. Ackerman, M.M.; Tang, X.; Guyot-Sionnest, P. Fast and sensitive colloidal quantum dot mid-wave infrared photodetectors. *ACS Nano* **2018**, *12*, 7264–7271. [CrossRef] [PubMed]
118. Tang, X.; Ackerman, M.M.; Shen, G.; Guyot-Sionnest, P.J. Towards infrared electronic eyes: Flexible colloidal quantum dot photovoltaic detectors enhanced by resonant cavity. *Small* **2019**, *15*, 1804920. [CrossRef] [PubMed]
119. Tang, X.; Lai, K.W.C. Graphene/HgTe quantum-dot photodetectors with gate-tunable infrared response. *ACS Appl. Nano Mater.* **2019**, *2*, 6701–6706. [CrossRef]
120. Tang, X.; Chen, M.; Kamath, A.; Ackerman, M.M.; Guyot-Sionnest, P.J. Colloidal quantum-dots/graphene/silicon dual-channel detection of visible light and short-wave infrared. *ACS Photonics* **2020**, *7*, 1117–1121. [CrossRef]
121. Dayeh, S.A.; Aplin, D.P.; Zhou, X.; Yu, P.K.; Yu, E.T.; Wang, D.J. High electron mobility InAs nanowire field-effect transistors. *Small* **2007**, *3*, 326–332. [CrossRef]
122. Ford, A.C.; Ho, J.C.; Chueh, Y.-L.; Tseng, Y.-C.; Fan, Z.; Guo, J.; Bokor, J.; Javey, A. Diameter-dependent electron mobility of InAs nanowires. *Nano Lett.* **2009**, *9*, 360–365. [CrossRef]
123. Takahashi, T.; Takei, K.; Adabi, E.; Fan, Z.; Niknejad, A.M.; Javey, A. Parallel array InAs nanowire transistors for mechanically bendable, ultrahigh frequency electronics. *ACS Nano* **2010**, *4*, 5855–5860. [CrossRef]
124. Meitei, P.; Alam, M.W.; Ngangbam, C.; Singh, N.K. Enhanced UV photodetection characteristics of annealed Gd2O3 nanorods. *Appl. Nanosci.* **2021**, *11*, 1437–1445. [CrossRef]
125. Geim, A.K.; Novoselov, K.S. The rise of graphene. *Nanosci. Technol.* **2009**, 11–19.
126. Partoens, B.; Peeters, F. From graphene to graphite: Electronic structure around the K point. *Phys. Rev. B* **2006**, *74*, 075404. [CrossRef]
127. Morozov, S.; Novoselov, K.; Schedin, F.; Jiang, D.; Firsov, A.; Geim, A. Two-dimensional electron and hole gases at the surface of graphite. *Phys. Rev. B* **2005**, *72*, 201401. [CrossRef]
128. Novoselov, K.S.; Geim, A.K.; Morozov, S.V.; Jiang, D.-E.; Zhang, Y.; Dubonos, S.V.; Grigorieva, I.V.; Firsov, A.A. Electric field effect in atomically thin carbon films. *Science* **2004**, *306*, 666–669. [CrossRef] [PubMed]
129. Peng, J.; Gao, W.; Gupta, B.K.; Liu, Z.; Romero-Aburto, R.; Ge, L.; Song, L.; Alemany, L.B.; Zhan, X.; Gao, G. Graphene quantum dots derived from carbon fibers. *Nano Lett.* **2012**, *12*, 844–849. [CrossRef]
130. Hai, X.; Feng, J.; Chen, X.; Wang, J.J. Tuning the optical properties of graphene quantum dots for biosensing and bioimaging. *J. Mater. Chem. B* **2018**, *6*, 3219–3234. [CrossRef]
131. Dong, Y.; Chen, C.; Zheng, X.; Gao, L.; Cui, Z.; Yang, H.; Guo, C.; Chi, Y.; Li, C.M. One-step and high yield simultaneous preparation of single-and multi-layer graphene quantum dots from CX-72 carbon black. *J. Mater. Chem.* **2012**, *22*, 8764–8766. [CrossRef]
132. Chen, X.; Ponraj, J.S.; Fan, D.; Zhang, H. An overview of the optical properties and applications of black phosphorus. *Nanoscale* **2020**, *12*, 3513–3534. [CrossRef]
133. Chen, H.; Huang, P.; Guo, D.; Xie, G.J.T. Anisotropic mechanical properties of black phosphorus nanoribbons. *J. Phys. Chem. C* **2016**, *120*, 29491–29497. [CrossRef]
134. Li, L.; Yu, Y.; Ye, G.J.; Ge, Q.; Ou, X.; Wu, H.; Feng, D.; Chen, X.H.; Zhang, Y. Black phosphorus field-effect transistors. *Nat. Nanotechnol.* **2014**, *9*, 372–377. [CrossRef]
135. Tran, V.; Soklaski, R.; Liang, Y.; Yang, L. Layer-controlled band gap and anisotropic excitons in few-layer black phosphorus. *Phys. Rev. B* **2014**, *89*, 235319. [CrossRef]
136. Qiao, J.; Kong, X.; Hu, Z.-X.; Yang, F.; Ji, W. High-mobility transport anisotropy and linear dichroism in few-layer black phosphorus. *Nat. Commun.* **2014**, *5*, 4475. [CrossRef] [PubMed]
137. Rao, C.E.N.E.R.; Sood, A.E.K.; Subrahmanyam, K.E.S.; Govindaraj, A. Graphene: The new two-dimensional nanomaterial. *J. Ger. Chem. Soc.* **2009**, *48*, 7752–7777.
138. Sidorov, A.N.; Yazdanpanah, M.M.; Jalilian, R.; Ouseph, P.; Cohn, R.; Sumanasekera, G. Electrostatic deposition of graphene. *Nanotechnology* **2007**, *18*, 135301. [CrossRef] [PubMed]
139. Berger, C.; Song, Z.; Li, T.; Li, X.; Ogbazghi, A.Y.; Feng, R.; Dai, Z.; Marchenkov, A.N.; Conrad, E.H.; First, P.N. Ultrathin epitaxial graphite: 2D electron gas properties and a route toward graphene-based nanoelectronics. *J. Phys. Chem. B* **2004**, *108*, 19912–19916. [CrossRef]
140. Ramachandran, V.; Brady, M.; Smith, A.; Feenstra, R.; Greve, D.J. Preparation of atomically flat surfaces on silicon carbide using hydrogen etching. *J. Electron. Mater.* **1998**, *27*, 308–312. [CrossRef]

141. Charrier, A.; Coati, A.; Argunova, T.; Thibaudau, F.; Garreau, Y.; Pinchaux, R.; Forbeaux, I.; Debever, J.-M.; Sauvage-Simkin, M.; Themlin, J.-M. Solid-state decomposition of silicon carbide for growing ultra-thin heteroepitaxial graphite films. *J. Appl. Phys.* **2002**, *92*, 2479–2484. [CrossRef]
142. Fogarassy, Z.; Dobrik, G.; Varga, L.K.; Biró, L.P.; Lábár, J.L. Growth of Ni layers on single crystal sapphire substrates. *Thin Solid Films* **2013**, *539*, 96–101. [CrossRef]
143. Fogarassy, Z.; Rümmeli, M.H.; Gorantla, S.; Bachmatiuk, A.; Dobrik, G.; Kamarás, K.; Biró, L.P.; Havancsák, K.; Lábár, J.L. Dominantly epitaxial growth of graphene on Ni (1 1 1) substrate. *Appl. Surf. Sci.* **2014**, *314*, 490–499. [CrossRef]
144. Subrahmanyam, K.; Panchakarla, L.; Govindaraj, A.; Rao, C.J.T. Simple method of preparing graphene flakes by an arc-discharge method. *J. Phys. Chem. C* **2009**, *113*, 4257–4259. [CrossRef]
145. Seshadri, R.; Govindaraj, A.; Aiyer, H.N.; Sen, R.; Subbanna, G.; Raju, A.; Rao, C.J. Investigations of carbon nanotubes. *Curr. Sci.* **1994**, 839–847.
146. Liu, H.; Du, Y.; Deng, Y.; Peide, D.Y. Semiconducting black phosphorus: Synthesis, transport properties and electronic applications. *Chem. Soc. Rev.* **2015**, *44*, 2732–2743. [CrossRef]
147. Maruyama, Y.; Suzuki, S.; Kobayashi, K.; Tanuma, S.J. Synthesis and some properties of black phosphorus single crystals. *Phys. B+C* **1981**, *105*, 99–102. [CrossRef]
148. Shirotani, I.J.M.C.; Crystals, L. Growth of large single crystals of black phosphorus at high pressures and temperatures, and its electrical properties. *Mol. Cryst. Liq. Cryst.* **1982**, *86*, 203–211. [CrossRef]
149. Lange, S.; Schmidt, P.; Nilges, T. Au3SnP7@ black phosphorus: An easy access to black phosphorus. *ChemInform* **2007**, *46*, 4028–4035.
150. Nilges, T.; Kersting, M.; Pfeifer, T.J. A fast low-pressure transport route to large black phosphorus single crystals. *J. Solid State Chem.* **2008**, *181*, 1707–1711. [CrossRef]
151. Köpf, M.; Eckstein, N.; Pfister, D.; Grotz, C.; Krüger, I.; Greiwe, M.; Hansen, T.; Kohlmann, H.; Nilges, T.J. Access and in situ growth of phosphorene-precursor black phosphorus. *J. Cryst. Growth* **2014**, *405*, 6–10. [CrossRef]
152. Bonaccorso, F.; Lombardo, A.; Hasan, T.; Sun, Z.; Colombo, L.; Ferrari, A.C. Production and processing of graphene and 2d crystals. *Mater. Today* **2012**, *15*, 564–589. [CrossRef]
153. Wang, J.; Liang, M.; Fang, Y.; Qiu, T.; Zhang, J.; Zhi, L.J. Rod-coating: Towards large-area fabrication of uniform reduced graphene oxide films for flexible touch screens. *Adv. Mater.* **2012**, *24*, 2874–2878. [CrossRef]
154. Akinwande, D.; Petrone, N.; Hone, J. Two-dimensional flexible nanoelectronics. *Nat. Commun.* **2014**, *5*, 5678. [CrossRef]
155. Han, T.-H.; Lee, Y.; Choi, M.-R.; Woo, S.-H.; Bae, S.-H.; Hong, B.H.; Ahn, J.-H.; Lee, T.-W. Extremely efficient flexible organic light-emitting diodes with modified graphene anode. *Nat. Photonics* **2012**, *6*, 105–110. [CrossRef]
156. Wu, Y.; Zou, X.; Sun, M.; Cao, Z.; Wang, X.; Huo, S.; Zhou, J.; Yang, Y.; Yu, X.; Kong, Y.J.; et al. 200 GHz maximum oscillation frequency in CVD graphene radio frequency transistors. *ACS Appl. Mater. Interfaces* **2016**, *8*, 25645–25649. [CrossRef] [PubMed]
157. Liu, M.; Yin, X.; Ulin-Avila, E.; Geng, B.; Zentgraf, T.; Ju, L.; Wang, F.; Zhang, X. A graphene-based broadband optical modulator. *Nature* **2011**, *474*, 64–67. [CrossRef] [PubMed]
158. Miao, X.; Tongay, S.; Petterson, M.K.; Berke, K.; Rinzler, A.G.; Appleton, B.R.; Hebard, A.F. High efficiency graphene solar cells by chemical doping. *Nano Lett.* **2012**, *12*, 2745–2750. [CrossRef] [PubMed]
159. Larki, F.; Abdi, Y.; Kameli, P.; Salamati, H. An effort towards full graphene photodetectors. *Photonic Sens.* **2022**, *12*, 31–67. [CrossRef]
160. Castellanos-Gomez, A.J.T. Black phosphorus: Narrow gap, wide applications. *J. Phys. Chem. Lett.* **2015**, *6*, 4280–4291. [CrossRef]
161. Liu, H.; Neal, A.T.; Zhu, Z.; Luo, Z.; Xu, X.; Tománek, D.; Ye, P.D. Phosphorene: An unexplored 2D semiconductor with a high hole mobility. *ACS Nano* **2014**, *8*, 4033–4041. [CrossRef]
162. Koenig, S.P.; Doganov, R.A.; Schmidt, H.; Castro Neto, A.; Özyilmaz, B. Electric field effect in ultrathin black phosphorus. *Appl. Phys. Lett.* **2014**, *104*, 103106. [CrossRef]
163. Du, H.; Lin, X.; Xu, Z.; Chu, D. Recent developments in black phosphorus transistors. *J. Mater. Chem. C* **2015**, *3*, 8760–8775. [CrossRef]
164. Eswaraiah, V.; Zeng, Q.; Long, Y.; Liu, Z. Black phosphorus nanosheets: Synthesis, characterization and applications. *Small* **2016**, *12*, 3480–3502. [CrossRef]
165. Zhu, W.; Yogeesh, M.N.; Yang, S.; Aldave, S.H.; Kim, J.-S.; Sonde, S.; Tao, L.; Lu, N.; Akinwande, D. Flexible black phosphorus ambipolar transistors, circuits and AM demodulator. *Nano Lett.* **2015**, *15*, 1883–1890. [CrossRef]

Review

X-ray Detectors Based on Halide Perovskite Materials

Yimei Tan [1], Ge Mu [1,*], Menglu Chen [1,2,3] and Xin Tang [1,2,3,*]

1 School of Optics and Photonics, Beijing Institute of Technology, Beijing 100081, China
2 Beijing Key Laboratory for Precision Optoelectronic Measurement Instrument and Technology, Beijing 100081, China
3 Yangtze Delta Region Academy of Beijing Institute of Technology, Jiaxing 314019, China
* Correspondence: 7520210145@bit.edu.cn (G.M.); xintang@bit.edu.cn (X.T.)

Abstract: Halide perovskite has remarkable optoelectronic properties, such as high atomic number, large carrier mobility-lifetime product, high X-ray attenuation coefficient, and simple and low-cost synthesis process, and has gradually developed into the next-generation X-ray detection materials. Halide perovskite-based X-ray detectors can improve the sensitivity and reduce the detectable X-ray dose, which is applied in imaging, nondestructive industrial inspection, security screening, and scientific research. In this article, we introduce the fabrication methods of halide perovskite film and the classification and progress of halide perovskite-based X-ray detectors. Finally, the existing challenges are discussed, and the possible directions for future applications are explored. We hope this review can stimulate the further improvement of perovskite-based X-ray detectors.

Keywords: halide perovskite; X-ray detector; film fabrication methods; sensitivity

Citation: Tan, Y.; Mu, G.; Chen, M.; Tang, X. X-ray Detectors Based on Halide Perovskite Materials. *Coatings* **2023**, *13*, 211. https://doi.org/10.3390/coatings13010211

Academic Editors: Alicia de Andrés and Torsten Brezesinski

Received: 11 December 2022
Revised: 9 January 2023
Accepted: 12 January 2023
Published: 16 January 2023

1. Introduction

X-ray detectors are commonly used in medical imaging, nondestructive industrial inspection, safety screening, and scientific research [1–3]. High sensitivity and high detection efficiency are believed to be critical figures of merit for X-ray detectors because weak X-ray signals can be detected, which greatly reduces the risk of medical examination [4–6]. The performance is closely related to the features of X-ray detection semiconductor materials including atomic number, charge carrier mobility, and carrier lifetime [7–9]. Various semiconductor materials have been utilized in X-ray detectors, such as silicon (Si), amorphous selenium (α-Se), germanium (Ge), and cadmium zinc telluride (CdZnTe). In addition, diamond is considered the elective material for X-ray detection and has a wide range of applications in the radiotherapy field mainly due to its "tissue equivalence" characteristic [10–13]. However, traditional materials such as α-Se, Ge, and CdZnTe suffer from many issues, including relatively small atomic numbers, the low attenuation coefficient of X-rays, and complex and costly fabrication processes [14–16]. Thus, it is necessary to explore new materials to replace the traditional materials for X-ray detectors.

Halide perovskites with a formula of ABX_3 (where A = $CH_3NH_3^+$ (MA^+), $HC(NH_2)_2^+$ (FA^+) and Cs^+; B = Pb^{2+} or Sn^{2+}; X = halides) are emerging candidates in X-ray detection. They have very high X-ray attenuation coefficients, a large carrier mobility-lifetime product ($\mu\tau$ product), a high atomic number, and a simple and low-cost synthesis process [17,18]. The summarization of the parameters of the materials for X-ray detectors is shown in Table 1. X-ray detectors employing organic–inorganic hybrid perovskite materials such as $MAPbI_3$ and $MAPbBr_3$ and all-inorganic perovskite materials such as $CsPbBr_3$ have been reported. According to the structural dimensionalities of halide perovskites, various types of materials with different characteristics of three-dimensional, low-dimensional, and three-dimensional/low-dimensional hybrid perovskite appear successively. Moreover, different techniques are used for the fabrication of high-quality halide perovskite films, which is the premise for achieving useful X-ray detectors. However, reviews that comprehensively summarize the fabrication methods of halide perovskite films and the features of halide perovskite-based X-ray detectors with different structural dimensionalities and different organic and inorganic components are relatively rare.

Table 1. The summarization of the parameters of the materials for X-ray detectors.

Material	Atomic Number	Density (g cm^{-3})	Band Gap (eV)	μτ Product (cm^2 V^{-1})	Resistivity (Ω cm)	Ref.
Si	14	2.33	1.12	>1	10^4	[19]
α-Se	34	4.3	2.1–2.2	10^{-7}	10^4–10^5	[20]
CdTe	48, 52	6.2	1.44	1.5×10^4	10^8–10^9	[21]
$Cd_{0.9}Zn_{0.1}Te$	48, 30, 52	5.78	1.57	10^{-2}	10^{11}	[21,22]
PbI_2	82, 53	6.2	2.3–2.6	10^{-5}	10^{13}	[20]
HgI_2	80, 53	6.4	2.13	10^{-4}	10^{13}	[23]
Ge	32	5.33	1.57	>1	50	[24]
$CsPbBr_3$	56.4	4.55	2.2–2.33	1.3×10^{-2}	8.5×10^9	[25]
$MAPbI_3$	35.6	4.3	1.5–1.6	10^{-4}–10^{-7}	-	[26]
$MAPbBr_3$	35, 82	3.45	2.2–2.3	-	-	[5]
Diamond	12	3.52	5.25–5.5	-	-	[10–13,20]

This review introduces in detail the fabrication methods of halide perovskite films and then summarizes the progress of research and development of halide perovskite-based X-ray detectors. Figure 1 shows the main fabrication methods and types of detectors introduced in this review. Finally, according to the progress of halide perovskite-based X-ray detectors, the existing problems are discussed, and the possible directions for future applications are explored.

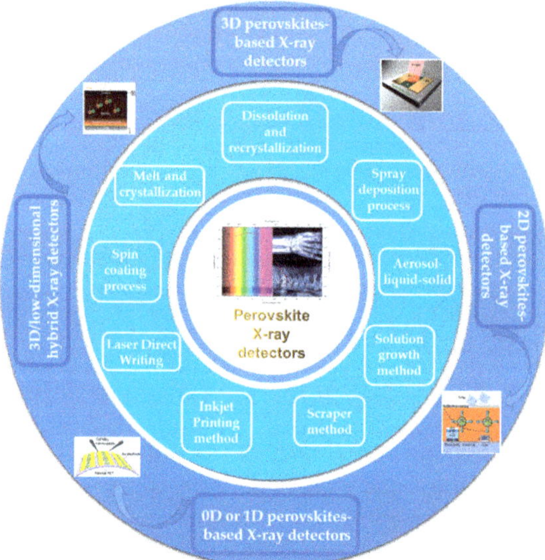

Figure 1. Progress in X-ray detectors based on halide perovskite materials.

2. Halide Perovskite Film Fabrication Methods

The quality of the film is the key factor to determine the performance of perovskite-based X-ray detectors. Fabrication methods are classified as a spin-coating process, dissolution and recrystallization method, spray deposition process, aerosol–liquid–solid method, solution growth method, scraper method, inkjet printing method, and laser direct writing and melt and crystallization. In the following, we describe these methods, introduce representative works, and analyze their features.

2.1. Spin-Coating Process

Spin coating is an early and widely used technology in perovskite film processing due to its simple and easy operation process. The process is mainly divided into two steps: dropping the solution on the cleaned substrate and accelerating the rotating plate to a required velocity. A uniform film is formed on the substrate because of the solution diffusion and solvent evaporation at high speed. The thickness of the film depends on the concentration, density, and spin velocity of the solution. In 2015, Yakunin et al. fabricated an X-ray detector based on methylammonium lead iodide perovskite ($CH_3NH_3PbI_3$) perovskite materials through the spin-coating process [27]. Uniform perovskite films deposited onto patterned electrode structures are accomplished through spin-coating technology (Figure 2a). Although the achievable film quality is high, the spin-coating process suffers from the difficulty of scaling up to large-area substrates. In addition, the rotational speed, solution concentration, etc., of different solutions need to be reexplored to obtain the desired thickness.

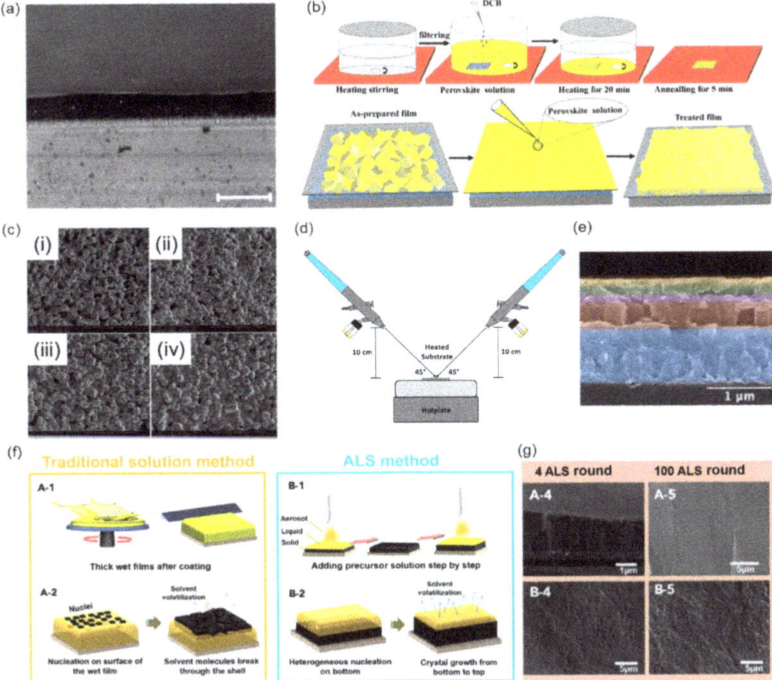

Figure 2. Spin-coating, dissolution and recrystallization, spray deposition, and aerosol–liquid–solid methods. (**a**) The cross-sectional optical microscopy of perovskite films and the scale bar is 100 μm. Reprinted with permission from Ref. [27]. 2015, Springer Nature. (**b**) $CsPbBr_3$ thick film preparation and repair process diagram. Reprinted with permission from Ref. [26]. 2019, Wiley. (**c**) SEM images of (i) the original $CsPbBr_3$ thick film, and the treated thick films with (ii) 5, (iii) 10, and (iv) 20 times dissolution and recrystallization. Reprinted with permission from Ref. [26]. 2019, Wiley. (**d**) Spraying device for depositing double-layer perovskite film. Reprinted with permission from Ref. [28]. 2020, Optica. (**e**) SEM cross-sections of different layers of perovskite solar cells. Reprinted with permission from Ref. [28]. 2020, Optica. (**f**) Comparison of traditional solution-based approach to film production (**left**) and the aerosol–liquid–solid method (**right**). This highlights the latter's unique ability to manufacture high-quality thick perovskite films. Reprinted with permission from Ref. [29]. 2021, Cell Press. (**g**) Cross-section (**upper**) and top-view SEM images (**lower**) of perovskite films fabricated by the aerosol–liquid–solid method after 4 (**left**) and 100 growth cycles (**right**). Reprinted with permission from Ref. [29]. 2021, Cell Press.

2.2. Dissolution and Recrystallization Method

The dissolution and recrystallization process is an efficient method for the preparation of high-quality perovskite films. During many dissolution and recrystallization processes, the sharp portion of the surface particles of the film is dissolved in the original perovskite solution. This dissolved material recrystallizes at the right temperature, filling the holes in the surface. After this repeated process, smooth and dense thick perovskite films can be obtained. In 2019, Zongyan Gou et al. ameliorated the film of the $CsPbBr_3$ microcrystalline after many dissolutions and recrystallizations, laying a good foundation for obtaining a high-performance X-ray detector (Figure 2b) [26]. The scanning electron microscope (SEM) images of the $CsPbBr_3$ films after repairing at different times are shown in Figure 2c. The surface of the original membrane is rough and full of holes. With the increase in dissolution and recrystallization times, the surface of the $CsPbBr_3$ film becomes microporous and smooth. This indicates that the rigid part of the $CsPbBr_3$ crystal on the surface of the film gradually dissolves into the original perovskite solution, leading to the filling of the holes.

2.3. Spray Deposition Process

The spray deposition process is one of the low-cost and large-scale production manufacturing methods for perovskite films (Figure 2d). The perovskite can be easily modified into ink to form a uniform perovskite film by a spray deposition process. In 2020, Koth Amratisha et al. developed the continuous spray deposition technique to realize the layer-by-layer stacking structure of different perovskite materials (Figure 2e) [28]. The perovskite film stability can be well maintained in a humid environment owing to the spray deposition preparation method. Sequential spray deposition technology opens up a new way for perovskite film stacking design and mass production under economical ambient conditions.

2.4. Aerosol–Liquid–Solid Method

The aerosol–liquid–solid method is a streamlined circulation from aerosol to liquid to solid, which can be used for the preparation of perovskite film. A suite of technological parameters including temperature, aerosol supply rate, and composition can be precisely controlled for aerosol–liquid–solid method technology. Due to surface tension and viscosity limitations, it is tough to deposit very thick wet films on the substrate through traditional solution methods such as spin-coating methods; however, the aerosol–liquid–solid method could achieve dense, high crystallinity, low defect density thick perovskite film around 1 mm (Figure 2f). In 2021, Wei Qian et al. demonstrated aerosol–liquid–solid methods to enable the continued growth of homogeneous perovskite films [29]. The thickness and grain size of the film increase gradually with the increase in growth time, as shown in Figure 2g.

2.5. Solution Growth Method

The solution growth method has the advantages of a low-cost, large-scale, and fast growth rate, which is suitable for the growth of perovskite single crystal. In 2016, Haotong Wei et al. successfully fabricated solution-grown $MAPbBr_3$ single crystals with extraordinary optoelectronic properties [30]. In 2020, Xin Wang et al. achieved the solution-processed epitaxial growth of $MAPbX_3$ organic–inorganic hybrid perovskite single crystal with different halide components and no lattice mismatch [31]. Figure 3a shows the SEM images of the cross-section and surface characterization of the solution-processed epitaxial layer, revealing the growth of a single-crystal film from a small island into a flat film.

2.6. Scraper Method

The scraper method is one of the crafts for producing films. In 1952, a scraper method patent was obtained by scraping water-based and non-water-based slurry onto moving plasterboard. Recently, the scraper method has been applied in the fabrication of perovskite films. When a constant relative motion is established between the blade and the substrate, the perovskite slurry is spread over the substrate to form a thin perovskite film. The scraper can run at a speed of several meters/per min. This operation mode is suitable for coating wet perovskite films with a thickness of tens of microns to hundreds of microns [32,33]. In 2017, Yong Churl Kim et al. used a scraper method to fabricate an 830 μm-thick polycrystalline $MAPbI_3$ layer [34]. In 2022, Mengling Xia et al. applied a combination of the scraper method and a soft pressure-assisted cryogenic solution treatment to make high-quality films for thin-film transistor (TFT) integration (Figure 3b) [35]. The method realizes the compatibility between the perovskite material and the TFT substrate and helps to achieve compact perovskite films with a smooth surface and passivated grain boundaries.

2.7. Inkjet Printing Method

Inkjet printing technology represents a highly established method for thin-film manufacturing with applications in electronics, optics, bioengineering, and other fields [36]. This technology has the advantages of low manufacturing costs, a simple and flexible process, and no mask plate and lithography, which can be widely used in perovskite solution printing [37,38]. Figure 3c illustrates the basic components of an inkjet printer including an ink chamber, conducting nozzle, substrate, and translational stage [39]. When a voltage is applied between the substrate and the nozzle, a curved liquid surface is formed at the nozzle under the action of the induced electric field force [40]. As the voltage gradually increases, small droplets are formed and ejected onto the substrate (Figure 3d) [41]. The movement of the nozzle and the mode of the applied voltage can control the pattern printed on the substrate [39–41]. In 2019, Jingying Liu et al. homogeneously prepared perovskite films on different nature substrates by the inexpensive inkjet printing method (Figure 3e) [42]. The cheap and simple inkjet printing method enables the large-scale manufacturing of multi-channel arrays of perovskite-based X-ray detectors.

2.8. Laser Direct Writing

Laser direct writing is a common method for preparing nanostructured graphics. Laser direct writing technology is widely used in all-inorganic halide perovskite colloidal quantum dots. In 2017, Chen Jun et al. proposed a rapid preparation method of perovskite colloidal quantum-dot film based on laser direct writing. It is simple, fast, and does not require a mask [43]. It includes three steps: spin-coating perovskite colloidal quantum dots, laser writing, and solvent washing (Figure 3f). Moreover, a large-scale (100 mm $\times$ 100 mm) perovskite colloidal quantum-dot patterning is demonstrated by the laser direct writing technology.

2.9. Melt and Crystallization

Perovskite has various forms such as nanocrystals, nanowires, and polycrystalline films, among which perovskite single crystal is the more stable form. Melt and crystallization are the preferred methods to obtain perovskite crystals. In this way, high-quality perovskite single crystals can be grown from the melt, which can meet the requirements of large diameters and cause less pollution during the synthesis process. In 2022, Andrii Kanak et al. studied the melting and crystallization process of all-organic perovskite $CsPbBr_3$, obtained massive single crystals from the melt, discussed the phase transition mechanism of the whole process, and proposed the two-stage melting mechanism of $CsPbBr_3$ perovskite. The effect of heating and cooling conditions on the crystallization process of large particle $CsPbBr_3$ was studied. This study is helpful to further understanding the crystal structure and crystallization mechanism of perovskite. It is important for understanding the growth and high-quality preparation of perovskite crystals and massive perovskite materials [44].

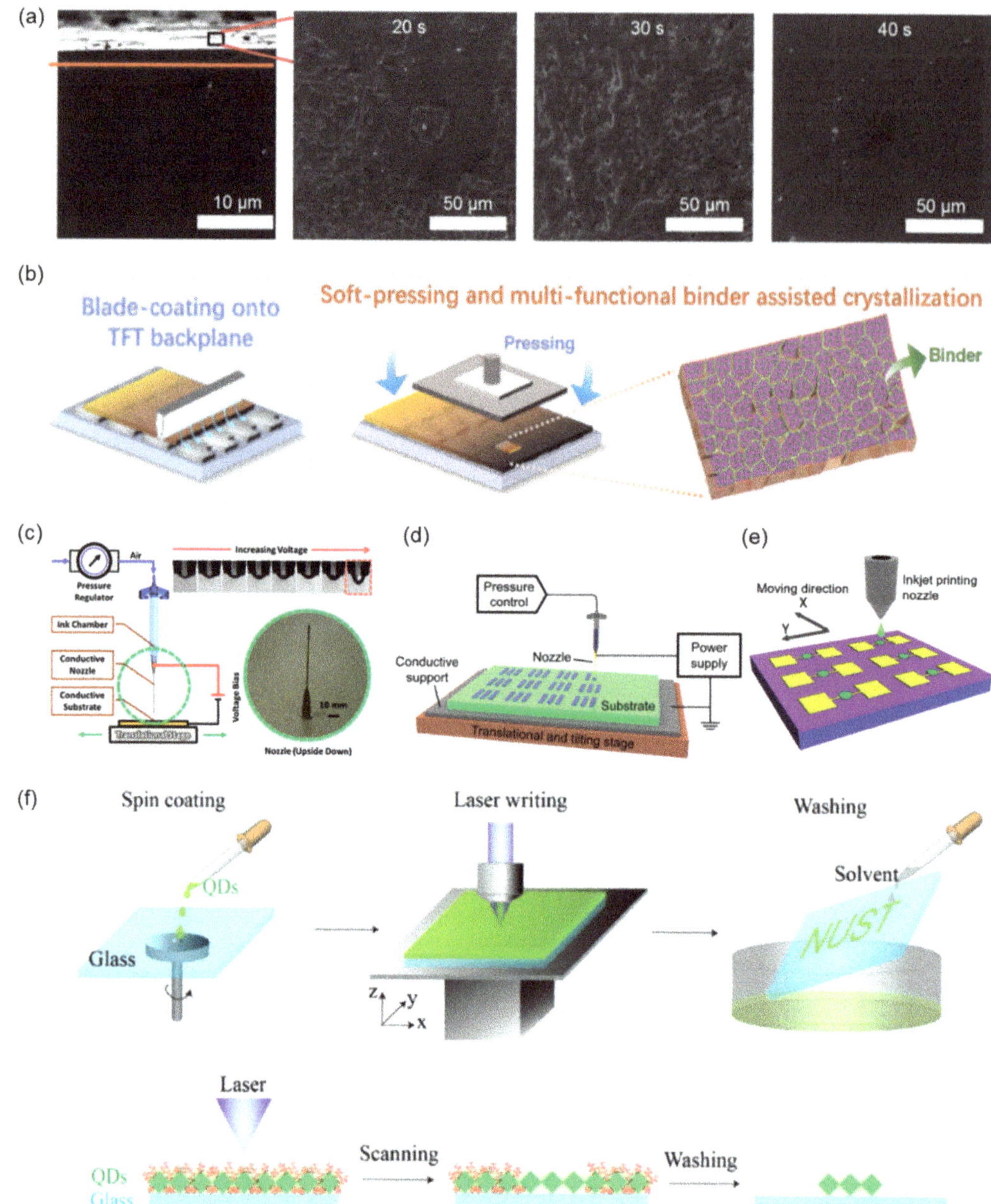

Figure 3. Solution growth, scraper, inkjet printing, and laser direct writing methods. (**a**) SEM photos of epitaxial layer cross-section and solution treatment at different growth times. Reprinted with permission from Ref. [31]. 2020, American Chemical Society. (**b**) Scraper method and a soft pressure-assisted cryogenic solution treatment. Reprinted with permission from Ref. [35]. 2022, Wiley. (**c**) Schematic of inkjet printer. Reprinted with permission from Ref. [39]. 2012, IOP Publishing. (**d**) Schematic illustration of inkjet printing system. Reprinted with permission from Ref. [41]. 2015, Royal Society of Chemistry. (**e**) Schematic diagram of a perovskite-based device manufactured by inkjet printing. Reprinted with permission from Ref. [42]. 2019, Wiley. (**f**) Schematic diagram of the modeling process and mechanism of perovskite colloidal quantum dots. Reprinted with permission from Ref. [43]. 2017, Wiley.

3. Halide Perovskite-Based X-ray Detector

Recently, there have been many high-impact review papers on introducing perovskite-based X-ray detectors. Zhizai Li et al. reviewed the current research progress on halide perovskite-based X-ray detectors [45]. Joydip Ghosh et al. summarized recent efforts on lead-free double perovskite-based X-ray detectors [46]. Haodi Wu et al. advised scalable fabrication of metal halide perovskite-based X-ray detectors [47].

Halide perovskite as an emerging X-ray detection material has a long carrier lifetime, large atomic numbers, and high X-ray attenuation coefficient [23,48]. An X-ray detector based on halide perovskite material possesses high detection sensitivity, which is crucial for detecting weak X-ray dose rates and greatly reducing the risk of medical examination [45]. Moreover, the synthesis process of halide perovskite is simple and low-cost as previously mentioned, which could form large-area flat panel detector arrays at low cost. Halide perovskite-based X-ray detector with high sensitivity and low cost has great application prospects in security, defense, medical imaging, industrial materials inspection, and nuclear power plants [49–51].

There are various structural dimensionalities of perovskites by controlling appropriate organic and inorganic components, such as 3D and low-dimensional (2D, 1D, 0D) perovskites. In recent years, various halide perovskite-based X-ray detectors have been developed by utilizing different dimensional perovskites. The three main types of halide perovskite-based X-ray detectors are described in detail below, including 3D, low, and 3D/low-dimensional mixed perovskite-based X-ray detectors. The summarization of the performance of X-ray detectors based on halide perovskite materials is shown in Table 2.

3.1. 3D Perovskites-Based X-ray Detectors

3D perovskites have a universal formula of ABX_3 (where A is a monovalent organic or inorganic cation, B is a divalent cation, and X is halide anion) and are composed of continuous corner-sharing metal halide $[BX_6]^{4-}$ octahedra [52]. According to the organic and inorganic components, 3D perovskites-based X-ray detectors are mainly divided into two representative types: organic–inorganic hybrid perovskites and all-inorganic perovskites-based X-ray detectors.

3.1.1. Organic–Inorganic Hybrid Perovskites-Based X-ray Detectors

Organic–inorganic hybrid perovskites have been shown to have good optoelectronic properties for X-ray detection, including a large appropriate band gap (1.6–3.0 eV) and a large $\mu\tau$ product on the order of ~10^{-2} cm^2 V^{-1} [53]. In 2017, Wei Wei et al. fabricated a $MAPbBr_3$ single-crystal X-ray detector and integrated it onto the Si substrates. This allows the electrical signal to be read directly from the Si (Figure 4a) [53]. The brominated (3-amino-propyl) triethoxy-silane (NH_3Br) end molecular layer is used to mechanically and electrically connect the $MAPbBr_3$ single crystals to the Si. No lattice matching with the Si substrate is required. The sensitivity of the $MAPbBr_3$ single-crystal X-ray detector is 2.1×10^4 $\mu C\ Gy_{air}{}^{-1}\ cm^{-2}$ under 8 keV X-ray radiation, which is more than 1000 times that of the commercially available amorphous α-Se detector. As shown in Figure 4b, it is possible to achieve X-ray imaging at a low X-ray (8 keV) dose rate of <0.1 $\mu Gy_{air}\ s^{-1}$.

Although halide perovskite X-ray detectors have had great success, the high concentration of Pb in perovskite poses a serious threat to human and biological systems due to its high water solubility. "Green" bi-element with the same electronic structure as the Pb attracts wide attention. In 2017, Weicheng Pan et al. demonstrated sensitive X-ray detectors using solution-processed double perovskite $Cs_2AgBiBr_6$ single crystals [54]. The $Cs_2AgBiBr_6$ single crystals use one Ag^+ and one Bi^{3+} to substitute two toxic Pb^{2+} in $CsPbBr_3$, which are very friendly to humans and biological systems (Figure 4c,d). In addition, after heat annealing and surface treatment, the Ag^+/Bi^{3+} disordering is eliminated and the crystal resistivity is enhanced. The $Cs_2AgBiBr_6$ single crystals exhibit a higher resistivity (10^9–10^{11} Ω cm) than $MAPbX_3$ (X = Cl, Br, I; 10^7–10^8 Ω cm). As a result, the $Cs_2AgBiBr_6$ single-crystal X-ray detectors possess a low minimum dose rate of 59.7 $nGy_{air}\ s^{-1}$ under

5 V and a high sensitivity of 8 μC Gy_{air}^{-1} cm^{-2} (1 V bias) to 30 keV X-ray photons and 105 μC Gy_{air}^{-1} cm^{-2} at 50 V (Figure 4e).

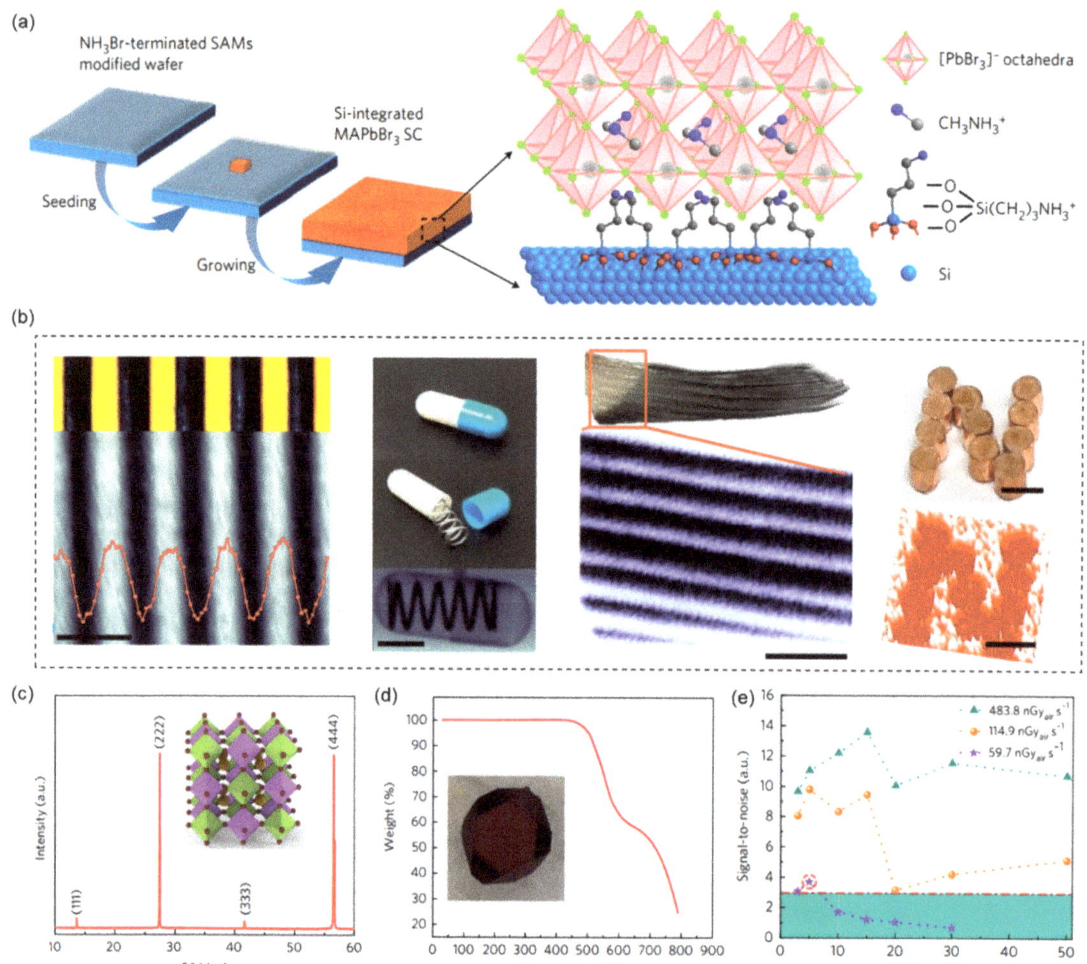

Figure 4. 3D organic–inorganic hybrid perovskites-based X-ray detectors. (**a**) Preparation diagram of Si-integrated $MAPbBr_3$ single crystal. Reprinted with permission from Ref. [53]. 2017, Springer Nature. (**b**) This device was used for the X-ray imaging in stacked glass coverslips; a stainless-steel plate with etched-through lines, a wrapped metal spring, a section of the tail fin, and an "N" copper sign gave a dose rate of 247 nGy_{air} s^{-1}. Reprinted with permission from Ref. [53]. 2017, Springer Nature. (**c**) X-ray diffraction of $Cs_2AgBiBr_6$ single crystal. Illustration: Crystal structure. The small wine ball represents Br^-, the large khaki ball represents Cs^+, and the light green and purple octahedra represent the $AgBr_6$ and $BiBr_6$ octahedra, respectively. Reprinted with permission from Ref. [54]. 2017, Springer Nature. (**d**) Thermogravimetric analysis of $Cs_2AgBiBr_6$. Included is a photograph of a solution-processed $Cs_2AgBiBr_6$ single crystal. Reprinted with permission from Ref. [54]. 2017, Springer Nature. (**e**) The signal-to-noise ratio of the device is calculated from the standard deviation of the X-ray photocurrent. The red dashed line represents a signal-to-noise ratio of 3, and thus the detection limit is 59.7 nGy_{air} s^{-1} at 5 V bias; it is represented by purple stars surrounded by red dotted circles. Reprinted with permission from Ref. [54]. 2017, Springer Nature.

3.1.2. All-Inorganic Perovskites-Based X-ray Detectors

Although the hybrid organic–inorganic perovskite-based X-ray detectors have realized outstanding performance, the stability of hybrid organic–inorganic perovskite is poor. This has an impact on their actual use, hindering their use in practical applications. The environmental conditions of their storage, fabrication, and device operation are very strict due to the extreme sensitivity to both oxygen and moisture. In addition, they are also unstable to light and heat due to the influence of organic groups. All-inorganic cesium halide lead perovskite ($CsPbX_3$) without organic components is considered to be the next generation of X-ray detector materials due to its superior stability to hybrid perovskite.

In 2018, Jin Hyuck Heo et al. prepared a $CsPbX_3$ perovskite nanocrystalline X-ray detector that is easy to commercialize and cost-effective (Figure 5a) [55]. The acquired X-ray detectors exhibit high stability over X-ray irradiation of 40 Gy_{air} s^{-1}. Moreover, the X-ray imaging possesses an excellent spatial resolution of 9.8 lp mm^{-1} at modulation transfer function (MTF) = 0.2 and 12.5–8.9 lp mm^{-1} for a linear line chart, which is higher than commercial terbium-doped gadolinium oxysulfide (GOS) detectors of spatial resolution = 6.2 lp mm^{-1} at MTF = 0.2 and 6.3 lp mm^{-1} for a linear line chart (Figure 5b,c).

In 2019, Zongyan Gou et al. fabricated X-ray detectors based on the $CsPbX_3$ microcrystal thick film. The multi-dissolution and recrystallization method is used to further improve the photoelectric performance of the detector (Figure 5d) [26]. The sensitivity of $CsPbX_3$-based X-ray detectors is 470 μC Gy_{air}^{-1} cm^{-2} at zero bias under a remarkably low dose rate (0.053 μGy_{air} s^{-1}), which is over 20 times higher than that of α-Se X-ray detectors working at a much higher field of 10 V μm^{-1} (Figure 5e,f).

3.2. *Low-Dimensional Perovskites-Based X-ray Detectors*

Although 3D perovskites-based X-ray detectors achieve high sensitivity, the phase transformation and instability of 3D perovskites limit their development. To address these issues of 3D perovskites, low-dimensional perovskites have been developed and high-performance low-dimensional perovskites-based X-ray detectors have been realized.

3.2.1. Organic–Inorganic Hybrid Perovskites-Based X-ray Detectors

$(NH_4)_3Bi_2I_9$ belongs to organic–inorganic hybrid perovskite-related materials $A_3M_2X_9$ (A = Cs, Rb, NH_4; M = Bi, Sb; X = Br, I). The $(NH_4)_3Bi_2I_9$ has a 2D layered structure where the BiI_6 octahedra corners share each other and are stacked in a close-packed fashion in the (001) plane (Figure 6a) [56]. In 2019, Renzhong Zhuang et al. verified the 2D $(NH_4)_3Bi_2I_9$ without toxic elements is a suitable material to obtain a high-performance X-ray detector [56]. Layered $(NH_4)_3Bi_2I_9$ grows easily in low-temperature solutions and can be cut along cleavage planes (Figure 6b,c). Two types of $(NH_4)_3Bi_2I_9$-based X-ray detectors with parallel and perpendicular device structures were proposed, as shown in Figure 6d. The devices were exposed to a source with X-ray photon energy up to 50 keV and with a peak intensity of 22 keV. The sensitivity of the parallel direction device is high at 8.2×10^3 μC Gy_{air}^{-1} cm^{-2}, which is higher than that of the perpendicular one (803 μC Gy_{air}^{-1} cm^{-2}). Moreover, the perpendicular direction device possesses a lower detection limit of 55 nGy_{air} s^{-1} compared to the parallel one of 210 nGy_{air} s^{-1} (Figure 6e).

3.2.2. All-Inorganic Perovskites-Based X-ray Detectors

0D perovskite quantum dots are ideal candidates for X-ray detectors and large-area flat or flexible panels with great application potential. $CsPbBr_3$ materials as a typical all inorganic perovskite possess superior stability and excellent optoelectronic properties. In 2019, Jingying Liu et al. synthesized 0D high-quality colloidal $CsPbBr_3$ perovskite quantum dots via the hot injection method at room temperature and demonstrated a flexible, printable $CsPbBr_3$-based X-ray detector (Figure 7a–c) [42]. Perovskite quantum-dot films are printed evenly on a variety of substrates using an inexpensive inkjet printing method to demonstrate large-scale manufacturing of X-ray detector arrays. The device was exposed to a synchrotron soft X-ray beamline with photon energies ranging from 0.1 to 2.5

keV. The detector has a sensitivity of 1450 $\mu Gy_{air}s^{-1}$ cm^{-2} at 0.0172 $mGy_{air}s^{-1}$ X-ray dose rate and a bias voltage of only 0.1 V, which is 70 times more sensitive than the α-Se device. Moreover, the $CsPbBr_3$ quantum-dot-based X-ray detectors possess outstanding flexibility and durability (Figure 7d–f).

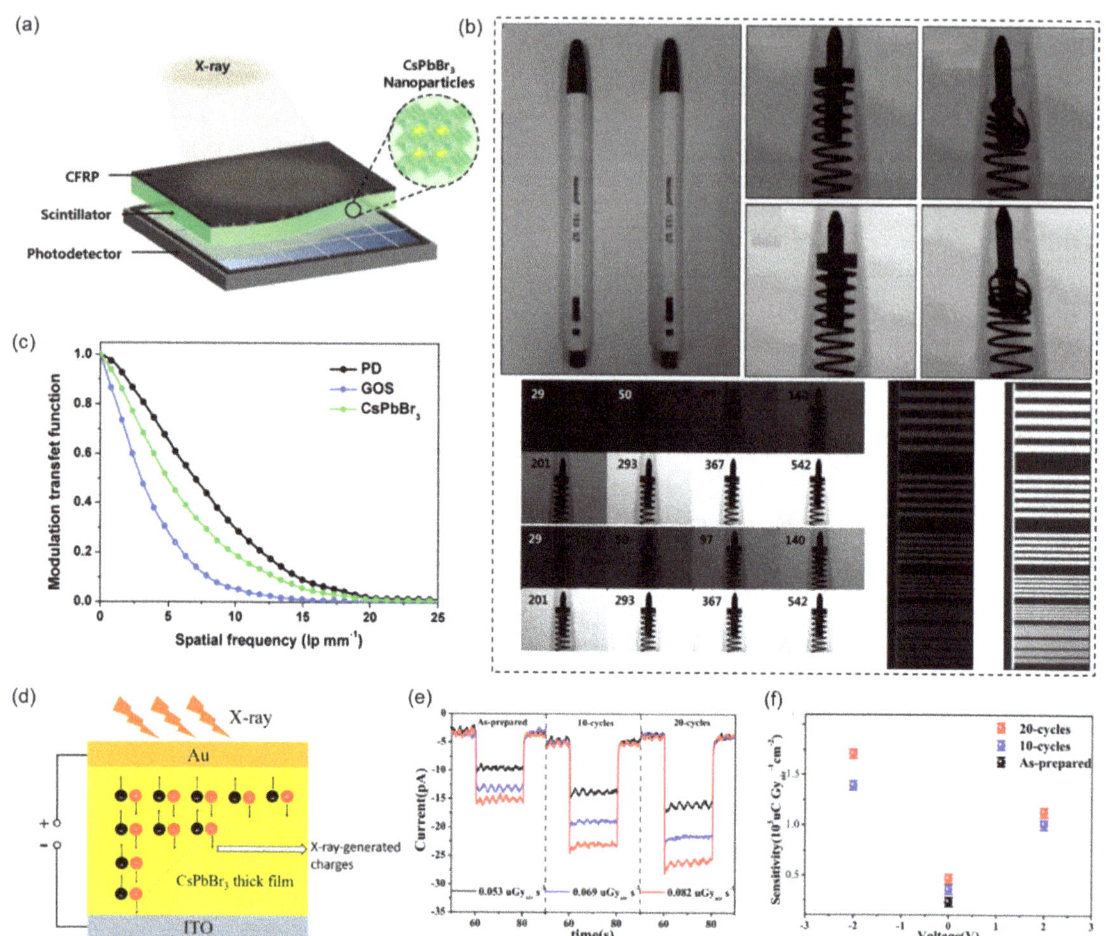

Figure 5. 3D all-inorganic perovskites-based X-ray detectors. (**a**) Schematic diagram of the structure of a $CsPbBr_3$ perovskite nanocrystalline scintillation X-ray detector. Reprinted with permission from Ref. [55]. 2018, Wiley. (**b**) Photographic and X-ray images for nondestructive inspection. Images of a normal ballpoint pen (**left**) and defective ballpoint pen (**right**) with identical appearance, ballpoint pen X-ray images from conventional GOS scintillator X-ray detectors, ballpoint pen X-ray images from $CsPbBr_3$ scintillator X-ray detectors, ballpoint pen X-ray images from GOS and $CsPbBr_3$ scintillator X-ray detectors with X-ray dose rates, line plots taken by GOS and $CsPbBr_3$ scintillation X-ray detectors. Reprinted with permission from Ref. [55]. 2018, Wiley. (**c**) MTF of the original silicon-based detector, $CsPbBr_3$, and conventional GOS scintillation X-ray detector. Reprinted with permission from Ref. [55]. 2018, Wiley. (**d**) Thick film radiation detector structure. The charge-producing region lies inside the thick film and is used for X-ray excitation. Reprinted with permission from Ref. [26]. 2019, Wiley. (**e**) Zero bias light response of raw and treated film at different dose rates. Reprinted with permission from Ref. [26]. 2019, Wiley. (**f**) The sensitivity of a device under different bias voltages and fixed emissivity (0.053 μGy_{air} s^{-1}). Reprinted with permission from Ref. [26]. 2019, Wiley.

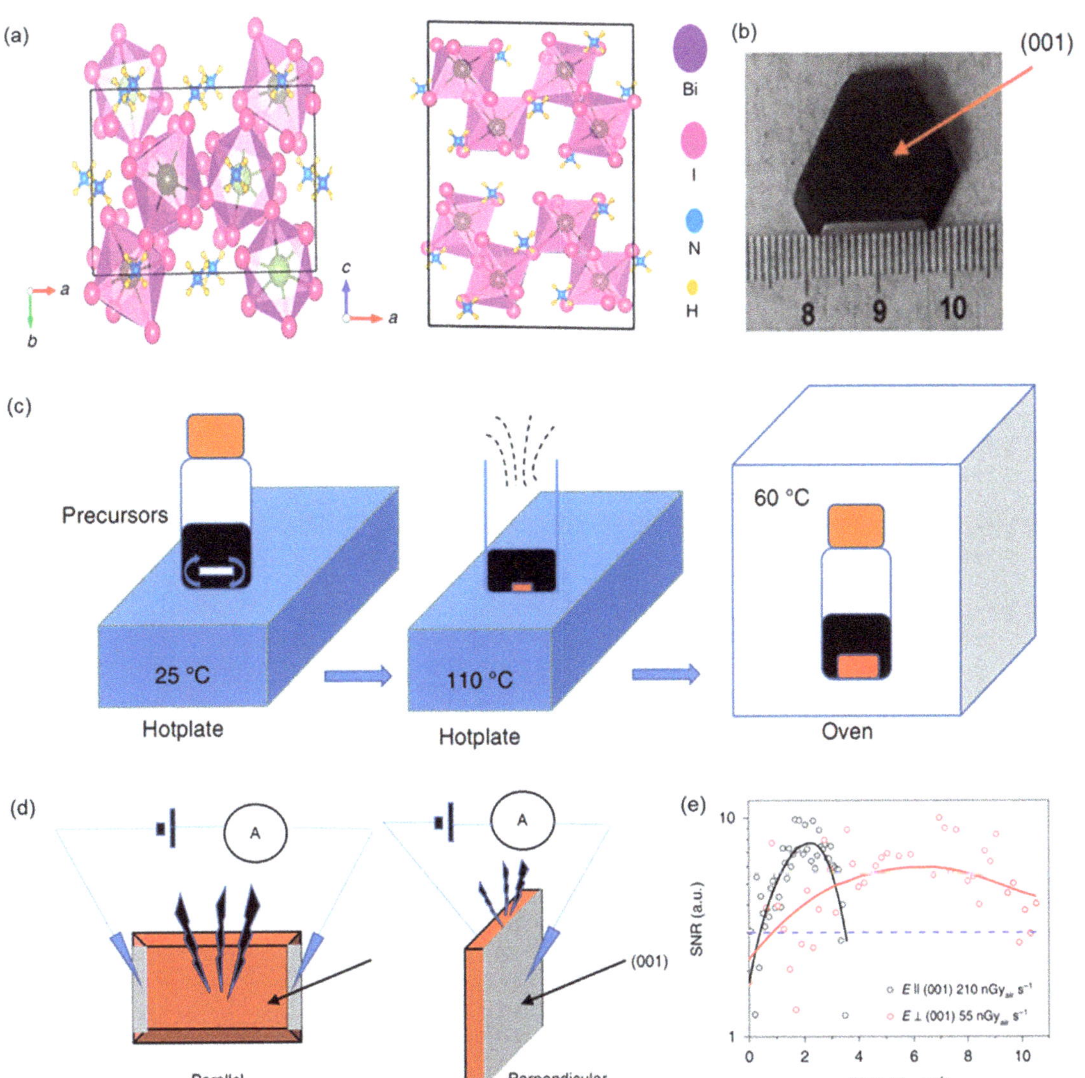

Figure 6. Low-dimensional organic–inorganic hybrid perovskites-based X-ray detectors. Reprinted with permission from Ref. [56]. 2019, Springer Nature. (**a**) Crystal structure of $(NH_4)_3Bi_2I_9$ along the c and b axes. (**b**) Photograph of a bulk $(NH_4)_3Bi_2I_9$ single crystal. (**c**) Procedure for crystal growth. (**d**) Illustration of parallel and perpendicular device structures. (**e**) The signal-to-noise ratio of the device in directions parallel to and perpendicular to the (001) surface. The dashed blue line indicates a signal-to-noise ratio of 3, so the detection limits are 210 $nGy_{air}\ s^{-1}$ for parallel devices and 55 $nGy_{air}\ s^{-1}$ for vertical devices.

3.3. Three-Dimensional/Low-Dimensional Hybrid Perovskites-Based X-ray Detectors

3D perovskites such as $MAPbI_3$, $MAPbBr_3$, and $CsPbBr_3$ are conducive to achieving high sensitivity due to their long carrier lifetime. However, baseline drifting problems arising from ion migration in the 3D perovskite material prevent accurate signal recording, hindering practical applications. Low-dimensional perovskites such as 2D $BDAPbI_4$ (BDA = $NH_3C_4H_8NH_3$) and $(NH_4)_3Bi_2I_9$ can effectively solve the baseline drifting problems. However, the sensitivity of low-dimensional perovskite is limited by its poor carrier transport performance. Thus, the combination of the 3D and low-dimensional perovskites could be an effective choice beyond the performance of X-ray detection with individual dimension perovskites.

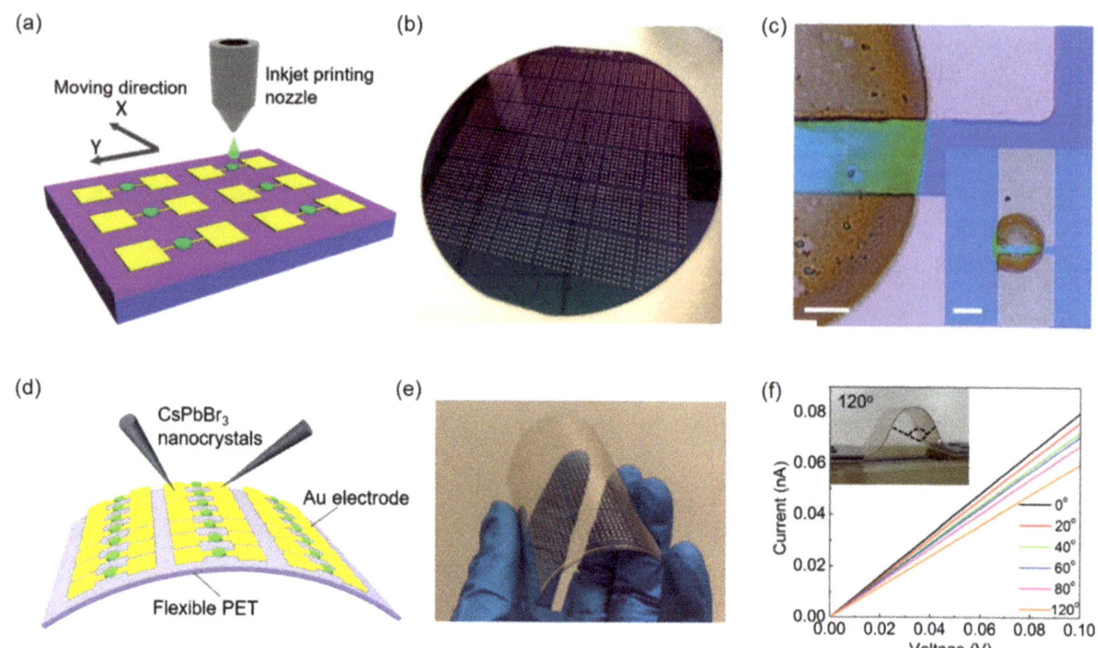

Figure 7. Low-dimensional all-inorganic perovskites-based X-ray detectors. Reprinted with permission from Ref. [42]. 2019, Wiley. (**a**) Schematic diagram of a perovskite-based device manufactured by inkjet printing. (**b**) Photography of an X-ray detector array on a 4-inch wafer. (**c**) Optical imaging equipment with the printing of quantum-dot film. Scale: 5 µm. Illustration: Low-power image. Scale: 50 µm. (**d**) Chemical reactions of flexible perovskite-based X-ray detector arrays on polyethylene terephthalate (PET) substrates. (**e**) Photography of flexible devices under bending. (**f**) Current and voltage curves of flexible device arrays at different bending angles under X-ray irradiation with 7.33 $mGy_{air}s^{-1}$ and 0.1 V bias voltages. Inset: Real device showing 120° bending.

In 2021, Xiuwen Xu et al. developed a double-layer perovskite film with a properly aligned energy level, where 2D $(PEA)_2MA_3Pb_4I_{13}$ (PEA = 2-phenylethylammonium, MA = methylammonium) is cascaded with vertically crystallized 3D $MAPbI_3$ (Figure 8a–d) [57]. Combining the fast carrier transport of the 3D layer and mitigated ion migration of the 2D layer, X-ray detectors provide high sensitivity and very stable baselines. Moreover, the 2D layer increases the film resistivity and enlarges the energy barrier for the hole injection without compromising carrier extraction (Figure 8e–g). As a result, the obtained double-layer perovskite detector exhibits a high sensitivity (1.95×10^4 µC $Gy_{air}{}^{-1}$ cm^{-2}) and a low detection limit (480 nGy_{air} s^{-1}) (Figure 8h). The double-layer perovskite detector exhibits excellent stability and a highly repeatable X-ray response when the X-ray is periodically on/off (Figure 8i). X-ray images of the printed circuit boards obtained with a double-layer perovskite detector with an active area of 3 mm × 3 mm are shown in Figure 8j.

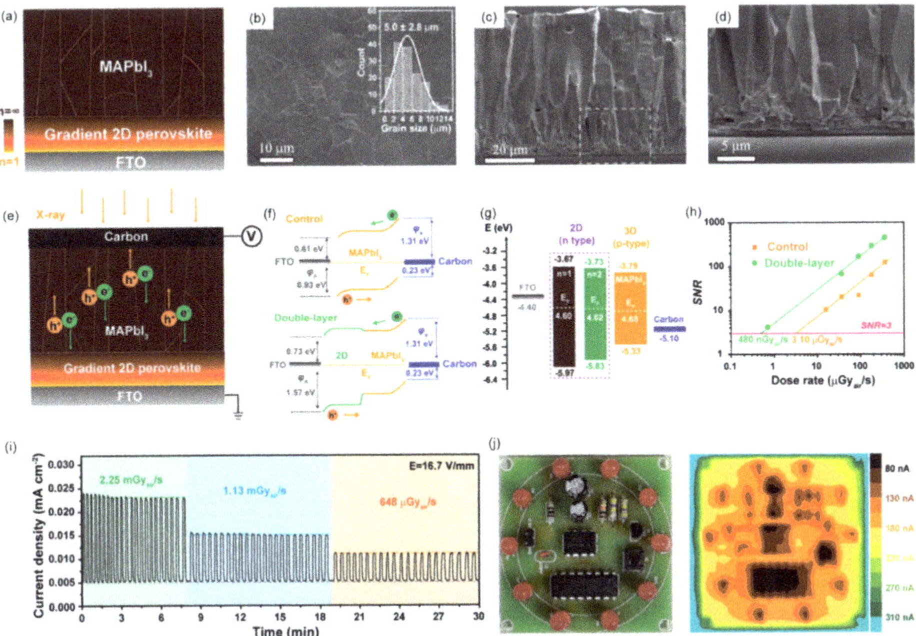

Figure 8. Three-dimensional/low-dimensional hybrid perovskites-based X-ray detectors. Reprinted with permission from Ref. [57]. 2021, Wiley. (**a**) Schematic diagram illustrating the structure of a double-layer perovskite thin film, where gradient 2D perovskite was prepared from a precursor solution with the nominal component $(PEA)_2MA_3Pb_4I_{13}$. (**b**–**d**) SEM image of double-layer perovskite film, shown in (**b**) is the grain size distribution. (**e**) Double-layer perovskite X-ray detector equipment configuration: Carbon electrodes were exposed to incident X-ray photons and negatively biased, whereas gradient 2D perovskite was prepared from a precursor solution with the designated component $(PEA)_2MA_3Pb_4I_{13}$. (**f**) Control the interfacial level alignment of a device made of a double-layer perovskite film. (**g**) Band edge position diagram of two-dimensional perovskite and three-dimensional $MAPbI_3$ with respect to fluorine-doped SnO_2 (FTO) and carbon. (**h**) The dose-rate-dependent signal-to-noise ratio of control and double-layer perovskite detectors. (**i**) Stability test of double-layer perovskite detector under different dose rate X-ray irradiation. (**j**) A digital photograph of the printed circuit board (size: 5.1 cm × 5.1 cm) and X-ray images of a printed circuit board obtained with a double-layer perovskite detector with an active area of 3 mm × 3 mm.

Table 2. The summarization of the performance of halide perovskite-based X-ray detectors.

Type	Material	Fabrication Method	Film Thickness (μm)	Sensor Area (mm^2)	Sensitivity ($\mu C\ Gy_{air}^{-1}\ cm^{-2}$)	Minimum Detectable Dose Rate ($\mu Gy_{air} s^{-1}$)	Ref.
3D perovskites-based X-ray detectors	$MAPbBr_3$	Solution-processed molecular bonding method	150	0.044	21,000	<0.1 (−1 V)	[53]
		Solution growth method	2000	-	80	0.5 (−0.1 V)	[30]
	$Cs_2AgBiBr_6$	Solution-processed method	1180	3.14	105 (50 V)	0.0597 (5 V)	[54]
		Laser direct writing	-	10,000	-	-	[43]
	$CsPbBr_3$	Dissolution and recrystallization method	18	50	470	0.053 (2 V)	[26]
		Melt and crystallization	-	-	-	-	[44]

Table 2. *Cont.*

Type	Material	Fabrication Method	Film Thickness (μm)	Sensor Area (mm^2)	Sensitivity ($\mu C\ Gy_{air}^{-1}\ cm^{-2}$)	Minimum Detectable Dose Rate ($\mu Gy_{air}s^{-1}$)	Ref.
	$CH_3NH_3PbI_3$	Spin-coating process	10–100	-	-	-	[27]
	$CsPbI_2Br$	Aerosol–liquid–solid method	One micrometer-hundreds of micrometers	10,000	148,000	0.28 (0.125 V μm^{-1})	[29]
	$MAPbI_3$	Scraper method	400	78,400	17,432 (500 V mm^{-1})	0.067 (5 V mm^{-1})	[35]
Low-dimensional perovskites-based X-ray detectors	$(NH_4)_3Bi_2I_9$	Solution growth method	4900 (parallel); 1500 (perpendicular)	6.75 (parallel); 22 (perpendicular)	8200 (parallel); 803 (perpendicular)	0.21 (parallel 1 V); 0.055 (perpendicular 10 V)	[56]
	$CsPbBr_3$	Inkjet printing method	0.6	0.06	1450	17.2 (0.1 V)	[42]
3D/low-dimensional hybrid perovskites-based X-ray detectors	2D $(PEA)_2MA_3Pb_4I_{13}$ /3D $MAPbI_3$	Spray deposition process	0.5	9	19,500	0.48 (<25 V mm^{-1})	[57]

4. Challenges and Perspectives

A variety of materials have been studied for X-ray detection because X-ray detectors have a wide range of applications including medical imaging, nondestructive industrial inspection, and safety screening. Diamond is considered the elective material for X-ray detection, mainly due to its "tissue equivalence" [10–13]. Recently, halide perovskites are emerging candidates in X-ray detection owing to the high X-ray attenuation coefficients, large $\mu\tau$ product, high atomic number, and the simple and low-cost synthesis process. Although halide perovskite-based X-ray detectors have made many remarkable signs of progress in the past few years, there are still some important problems that need further consideration to meet application requirements. The existing challenges are discussed, and the possible directions for future applications are explored in the following.

Firstly, compared with traditional materials such as Si, α-Se, and Ge, the stability of perovskite materials is still relatively poor, which seriously restricts their practical application. Therefore, more attention should be paid to improving the stability of perovskite materials in future research.

Secondly, the thickness of perovskite films with hundreds of micrometers is the prerequisite for complete X-ray attenuation. However, the fabrication of a thick perovskite film with good uniformity is still a challenge. Moreover, the cost of preparation of perovskite-based X-ray detectors should be further reduced. Hence, developing new preparation methods or optimizing the original methods is the future development trend.

Thirdly, the sensitivity and minimum detectable X-ray dose rate of perovskite-based X-ray detectors are not good enough for practical use. It is necessary to develop high-performance perovskite-based X-ray detectors with high sensitivity and low detection limits.

Finally, the development of large-area flexible next-generation X-ray detection and imaging technology based on perovskite is the future research direction.

Author Contributions: Conceptualization, X.T.; investigation, Y.T. and G.M.; writing—original draft preparation, Y.T. and G.M.; writing—review and editing, Y.T., G.M. and X.T.; supervision, X.T.; project administration, X.T.; funding acquisition, X.T., G.M. and M.C. All authors have read and agreed to the published version of the manuscript.

Funding: This work was funded by the National Key R&D Program of China (2021YFA0717600), the National Natural Science Foundation of China (NSFC No. 62035004 and NSFC No. 62105022) and the China Postdoctoral Science Foundation (2022M710396). X.T. is sponsored by the Young Elite Scientists Sponsorship Program by CAST (No. YESS20200163).

Institutional Review Board Statement: Not applicable.

Informed Consent Statement: Not applicable.

Data Availability Statement: Not applicable.

Conflicts of Interest: The authors declare no conflict of interest.

References

1. Kasap, S.; Kabir, M.Z.; Rowlands, J. Recent advances in X-ray photoconductors for direct conversion X-ray image detectors. *Curr. Appl. Phys.* **2006**, *6*, 288–292. [CrossRef]
2. Long, M.; Zhang, T.; Xu, W.; Zeng, X.; Xie, F.; Li, Q.; Chen, Z.; Zhou, F.; Wong, K.S.; Yan, K. Large-grain formamidinium PbI3–xBrx for high-performance perovskite solar cells via intermediate halide exchange. *Adv. Energy Mater.* **2017**, *7*, 1601882. [CrossRef]
3. Su, Y.; Ma, W.; Yang, Y.M. Perovskite semiconductors for direct X-ray detection and imaging. *J. Semicond.* **2020**, *41*, 51204. [CrossRef]
4. Mescher, H.; Schackmar, F.; Eggers, H.; Abzieher, T.; Zuber, M.; Hamann, E.; Baumbach, T.; Richards, B.S.; Hernandez-Sosa, G.; Paetzold, U.W. Flexible inkjet-printed triple cation perovskite X-ray detectors. *ACS Appl. Mater. Interfaces* **2020**, *12*, 15774–15784. [CrossRef] [PubMed]
5. Pan, W.; Yang, B.; Niu, G.; Xue, K.H.; Du, X.; Yin, L.; Zhang, M.; Wu, H.; Miao, X.S.; Tang, J. Hot-Pressed CsPbBr3 Quasi-Monocrystalline Film for Sensitive Direct X-ray Detection. *Adv. Mater.* **2019**, *31*, 1904405. [CrossRef]
6. Basiricò, L.; Ciavatti, A.; Fraboni, B. Solution-Grown Organic and Perovskite X-Ray Detectors: A New Paradigm for the Direct Detection of Ionizing Radiation. *Adv. Mater. Technol.* **2021**, *6*, 2000475. [CrossRef]
7. Sun, C.; Shen, X.; Zhang, Y.; Wang, Y.; Chen, X.; Ji, C.; Shen, H.; Shi, H.; Wang, Y.; William, W.Y. Highly luminescent, stable, transparent and flexible perovskite quantum dot gels towards light-emitting diodes. *Nanotechnology* **2017**, *28*, 365601. [CrossRef]
8. Wang, W.; Ma, Y.; Qi, L. High-performance photodetectors based on organometal halide perovskite nanonets. *Adv. Funct. Mater.* **2017**, *27*, 1603653. [CrossRef]
9. Hu, H.; Niu, G.; Zheng, Z.; Xu, L.; Liu, L.; Tang, J. Perovskite semiconductors for ionizing radiation detection. *EcoMat* **2022**, *4*, e12258. [CrossRef]
10. Pettinato, S.; Girolami, M.; Olivieri, R.; Stravato, A.; Caruso, C.; Salvatori, S. A diamond-based dose-per-pulse x-ray detector for radiation therapy. *Materials* **2021**, *14*, 5203. [CrossRef]
11. Pettinato, S.; Girolami, M.; Olivieri, R.; Stravato, A.; Caruso, C.; Salvatori, S. Time-resolved dosimetry of pulsed photon beams for radiotherapy based on diamond detector. *IEEE Sens. J.* **2022**, *22*, 12348–12356. [CrossRef]
12. Talamonti, C.; Kanxheri, K.; Pallotta, S.; Servoli, L. Diamond Detectors for Radiotherapy X-Ray Small Beam Dosimetry. *Front. Phys.* **2021**, *9*, 632299. [CrossRef]
13. Marinelli, M.; Felici, G.; Galante, F.; Gasparini, A.; Giuliano, L.; Heinrich, S.; Pacitti, M.; Prestopino, G.; Vanreusel, V.; Verellen, D. Design, realization, and characterization of a novel diamond detector prototype for FLASH radiotherapy dosimetry. *Med. Phys.* **2022**, *49*, 1902–1910. [CrossRef] [PubMed]
14. Grancini, G.; Roldán-Carmona, C.; Zimmermann, I.; Mosconi, E.; Lee, X.; Martineau, D.; Narbey, S.; Oswald, F.; De Angelis, F.; Graetzel, M. One-Year stable perovskite solar cells by 2D/3D interface engineering. *Nat. Commun.* **2017**, *8*, 1–8. [CrossRef]
15. Bakr, Z.H.; Wali, Q.; Fakharuddin, A.; Schmidt-Mende, L.; Brown, T.M.; Jose, R. Advances in hole transport materials engineering for stable and efficient perovskite solar cells. *Nano Energy* **2017**, *34*, 271–305. [CrossRef]
16. Rao, H.-S.; Chen, B.-X.; Wang, X.-D.; Kuang, D.-B.; Su, C.-Y. A micron-scale laminar MAPbBr 3 single crystal for an efficient and stable perovskite solar cell. *Chem. Commun.* **2017**, *53*, 5163–5166. [CrossRef]
17. Haruta, Y.; Kawakami, M.; Nakano, Y.; Kundu, S.; Wada, S.; Ikenoue, T.; Miyake, M.; Hirato, T.; Saidaminov, M.I. Scalable Fabrication of Metal Halide Perovskites for Direct X-ray Flat-Panel Detectors: A Perspective. *Chem. Mater.* **2022**, *34*, 5323–5333. [CrossRef]
18. Büchele, P.; Richter, M.; Tedde, S.F.; Matt, G.J.; Ankah, G.N.; Fischer, R.; Biele, M.; Metzger, W.; Lilliu, S.; Bikondoa, O. X-ray imaging with scintillator-sensitized hybrid organic photodetectors. *Nat. Photonics* **2015**, *9*, 843–848. [CrossRef]
19. Luo, Z.; G Moch, J.; S Johnson, S.; Chon Chen, C. A review on X-ray detection using nanomaterials. *Curr. Nanosci.* **2017**, *13*, 364–372. [CrossRef]
20. Owens, A. Semiconductor materials and radiation detection. *J. Synchrotron Radiat.* **2006**, *13*, 143–150. [CrossRef]
21. Bellazzini, R.; Spandre, G.; Brez, A.; Minuti, M.; Pinchera, M.; Mozzo, P. Chromatic X-ray imaging with a fine pitch CdTe sensor coupled to a large area photon counting pixel ASIC. *J. Instrum.* **2013**, *8*, C02028. [CrossRef]
22. Guo, J.; Xu, Y.; Yang, W.; Zhang, B.; Dong, J.; Jie, W.; Kanatzidis, M.G. Morphology of X-ray detector Cs 2 TeI 6 perovskite thick films grown by electrospray method. *J. Mater. Chem. C* **2019**, *7*, 8712–8719. [CrossRef]
23. Wei, H.; Huang, J. Halide lead perovskites for ionizing radiation detection. *Nat. Commun.* **2019**, *10*, 1–12. [CrossRef] [PubMed]
24. Kabir, M.Z.; Kasap, S. Photoconductors for x-ray image detectors. In *Springer Handbook of Electronic and Photonic Materials*; Springer: Berlin/Heidelberg, Germany, 2017; p. 1.
25. Matt, G.J.; Levchuk, I.; Knüttel, J.; Dallmann, J.; Osvet, A.; Sytnyk, M.; Tang, X.; Elia, J.; Hock, R.; Heiss, W. Sensitive Direct Converting X-Ray Detectors Utilizing Crystalline CsPbBr3 Perovskite Films Fabricated via Scalable Melt Processing. *Adv. Mater. Interfaces* **2020**, *7*, 1901575. [CrossRef]

26. Gou, Z.; Huanglong, S.; Ke, W.; Sun, H.; Tian, H.; Gao, X.; Zhu, X.; Yang, D.; Wangyang, P. Self-Powered X-Ray Detector Based on All-Inorganic Perovskite Thick Film with High Sensitivity Under Low Dose Rate. *Phys. Status Solidi (RRL)–Rapid Res. Lett.* **2019**, *13*, 1900094. [CrossRef]
27. Yakunin, S.; Sytnyk, M.; Kriegner, D.; Shrestha, S.; Richter, M.; Matt, G.J.; Azimi, H.; Brabec, C.J.; Stangl, J.; Kovalenko, M.V. Detection of X-ray photons by solution-processed lead halide perovskites. *Nat. Photonics* **2015**, *9*, 444–449. [CrossRef]
28. Amratisha, K.; Ponchai, J.; Kaewurai, P.; Pansa-Ngat, P.; Pinsuwan, K.; Kumnorkaew, P.; Ruankham, P.; Kanjanaboos, P. Layer-by-layer spray coating of a stacked perovskite absorber for perovskite solar cells with better performance and stability under a humid environment. *Opt. Mater. Express* **2020**, *10*, 1497–1508. [CrossRef]
29. Qian, W.; Xu, X.; Wang, J.; Xu, Y.; Chen, J.; Ge, Y.; Chen, J.; Xiao, S.; Yang, S. An aerosol-liquid-solid process for the general synthesis of halide perovskite thick films for direct-conversion X-ray detectors. *Matter* **2021**, *4*, 942–954. [CrossRef]
30. Wei, H.; Fang, Y.; Mulligan, P.; Chuirazzi, W.; Fang, H.-H.; Wang, C.; Ecker, B.R.; Gao, Y.; Loi, M.A.; Cao, L. Sensitive X-ray detectors made of methylammonium lead tribromide perovskite single crystals. *Nat. Photonics* **2016**, *10*, 333–339. [CrossRef]
31. Wang, X.; Li, Y.; Xu, Y.; Pan, Y.; Zhu, C.; Zhu, D.; Wu, Y.; Li, G.; Zhang, Q.; Li, Q. Solution-processed halide perovskite single crystals with intrinsic compositional gradients for X-ray detection. *Chem. Mater.* **2020**, *32*, 4973–4983. [CrossRef]
32. Uhland, S.A.; Holman, R.K.; Morissette, S.; Cima, M.J.; Sachs, E.M. Strength of green ceramics with low binder content. *J. Am. Ceram. Soc.* **2001**, *84*, 2809–2818. [CrossRef]
33. Baklouti, S.; Bouaziz, J.; Chartier, T.; Baumard, J.-F. Binder burnout and evolution of the mechanical strength of dry-pressed ceramics containing poly (vinyl alcohol). *J. Eur. Ceram. Soc.* **2001**, *21*, 1087–1092. [CrossRef]
34. Kim, Y.C.; Kim, K.H.; Son, D.-Y.; Jeong, D.-N.; Seo, J.-Y.; Choi, Y.S.; Han, I.T.; Lee, S.Y.; Park, N.-G. Printable organometallic perovskite enables large-area, low-dose X-ray imaging. *Nature* **2017**, *550*, 87–91. [CrossRef] [PubMed]
35. Xia, M.; Song, Z.; Wu, H.; Du, X.; He, X.; Pang, J.; Luo, H.; Jin, L.; Li, G.; Niu, G. Compact and Large-Area Perovskite Films Achieved via Soft-Pressing and Multi-Functional Polymerizable Binder for Flat-Panel X-Ray Imager. *Adv. Funct. Mater.* **2022**, *32*, 2110729. [CrossRef]
36. Onses, M.S.; Sutanto, E.; Ferreira, P.M.; Alleyne, A.G.; Rogers, J.A. Mechanisms, capabilities, and applications of high-resolution electrohydrodynamic jet printing. *Small* **2015**, *11*, 4237–4266. [CrossRef]
37. Chen, D.; Liang, J.; Pei, Q. Flexible and stretchable electrodes for next generation polymer electronics: A review. *Sci. China Chem.* **2016**, *59*, 659–671. [CrossRef]
38. Park, H.-G.; Byun, S.-U.; Jeong, H.-C.; Lee, J.-W.; Seo, D.-S. Photoreactive spacer prepared using electrohydrodynamic printing for application in a liquid crystal device. *ECS Solid State Lett.* **2013**, *2*, R52. [CrossRef]
39. Sutanto, E.; Shigeta, K.; Kim, Y.; Graf, P.; Hoelzle, D.; Barton, K.; Alleyne, A.; Ferreira, P.; Rogers, J.A. A multimaterial electrohydrodynamic jet (E-jet) printing system. *J. Micromech. Microeng.* **2012**, *22*, 045008. [CrossRef]
40. de Gans, B.J.; Schubert, U.S. Inkjet printing of polymer micro-arrays and libraries: Instrumentation, requirements, and perspectives. *Macromol. Rapid Commun.* **2003**, *24*, 659–666. [CrossRef]
41. Kim, K.; Kim, G.; Lee, B.R.; Ji, S.; Kim, S.-Y.; An, B.W.; Song, M.H.; Park, J.-U. High-resolution electrohydrodynamic jet printing of small-molecule organic light-emitting diodes. *Nanoscale* **2015**, *7*, 13410–13415. [CrossRef]
42. Liu, J.; Shabbir, B.; Wang, C.; Wan, T.; Ou, Q.; Yu, P.; Tadich, A.; Jiao, X.; Chu, D.; Qi, D. Flexible, printable soft-X-ray detectors based on all-inorganic perovskite quantum dots. *Adv. Mater.* **2019**, *31*, 1901644. [CrossRef] [PubMed]
43. Chen, J.; Wu, Y.; Li, X.; Cao, F.; Gu, Y.; Liu, K.; Liu, X.; Dong, Y.; Ji, J.; Zeng, H. Simple and fast patterning process by laser direct writing for perovskite quantum dots. *Adv. Mater. Technol.* **2017**, *2*, 1700132. [CrossRef]
44. Kanak, A.; Kopach, O.; Kanak, L.; Levchuk, I.; Isaiev, M.; Brabec, C.J.; Fochuk, P.; Khalavka, Y. Melting and Crystallization Features of CsPbBr3 Perovskite. *Cryst. Growth Des.* **2022**, *22*, 4115–4121. [CrossRef]
45. Li, Z.; Zhou, F.; Yao, H.; Ci, Z.; Yang, Z.; Jin, Z. Halide perovskites for high-performance X-ray detector. *Mater. Today* **2021**, *48*, 155–175. [CrossRef]
46. Ghosh, J.; Sellin, P.J.; Giri, P.K. Recent advances in lead-free double perovskites for x-ray and photodetection. *Nanotechnology* **2022**, *33*, 312001. [CrossRef]
47. Wu, H.; Ge, Y.; Niu, G.; Tang, J. Metal halide perovskites for X-ray detection and imaging. *Matter* **2021**, *4*, 144–163. [CrossRef]
48. Sytnyk, M.; Deumel, S.; Tedde, S.F.; Matt, G.J.; Heiss, W. A perspective on the bright future of metal halide perovskites for X-ray detection. *Appl. Phys. Lett.* **2019**, *115*, 190501. [CrossRef]
49. Kasap, S.; Frey, J.B.; Belev, G.; Tousignant, O.; Mani, H.; Greenspan, J.; Laperriere, L.; Bubon, O.; Reznik, A.; DeCrescenzo, G. Amorphous and polycrystalline photoconductors for direct conversion flat panel X-ray image sensors. *Sensors* **2011**, *11*, 5112–5157. [CrossRef]
50. Yaffe, M.; Rowlands, J. X-ray detectors for digital radiography. *Phys. Med. Biol.* **1997**, *42*, 1. [CrossRef]
51. Tegze, M.; Faigel, G. X-ray holography with atomic resolution. *Nature* **1996**, *380*, 49–51. [CrossRef]
52. Sun, S.; Lu, M.; Gao, X.; Shi, Z.; Bai, X.; Yu, W.W.; Zhang, Y. 0D perovskites: Unique properties, synthesis, and their applications. *Adv. Sci.* **2021**, *8*, 2102689. [CrossRef] [PubMed]
53. Wei, W.; Zhang, Y.; Xu, Q.; Wei, H.; Fang, Y.; Wang, Q.; Deng, Y.; Li, T.; Gruverman, A.; Cao, L. Monolithic integration of hybrid perovskite single crystals with heterogenous substrate for highly sensitive X-ray imaging. *Nat. Photonics* **2017**, *11*, 315–321. [CrossRef]

54. Pan, W.; Wu, H.; Luo, J.; Deng, Z.; Ge, C.; Chen, C.; Jiang, X.; Yin, W.-J.; Niu, G.; Zhu, L. Cs2AgBiBr6 single-crystal X-ray detectors with a low detection limit. *Nat. Photonics* **2017**, *11*, 726–732. [CrossRef]
55. Heo, J.H.; Shin, D.H.; Park, J.K.; Kim, D.H.; Lee, S.J.; Im, S.H. High-performance next-generation perovskite nanocrystal scintillator for nondestructive X-ray imaging. *Adv. Mater.* **2018**, *30*, 1801743. [CrossRef] [PubMed]
56. Zhuang, R.; Wang, X.; Ma, W.; Wu, Y.; Chen, X.; Tang, L.; Zhu, H.; Liu, J.; Wu, L.; Zhou, W. Highly sensitive X-ray detector made of layered perovskite-like (NH4) 3Bi2I9 single crystal with anisotropic response. *Nat. Photonics* **2019**, *13*, 602–608. [CrossRef]
57. Xu, X.; Qian, W.; Wang, J.; Yang, J.; Chen, J.; Xiao, S.; Ge, Y.; Yang, S. Sequential Growth of 2D/3D Double-Layer Perovskite Films with Superior X-Ray Detection Performance. *Adv. Sci.* **2021**, *8*, 2102730. [CrossRef]

 coatings

Review

Multi-Color Light-Emitting Diodes

Su Ma [1,†], Yawei Qi [1,†], Ge Mu [1,*], Menglu Chen [1,2,3] and Xin Tang [1,2,3,*]

1 School of Optics and Photonics, Beijing Institute of Technology, Beijing 100081, China
2 Beijing Key Laboratory for Precision Optoelectronic Measurement Instrument and Technology, Beijing 100081, China
3 Yangtze Delta Region Academy of Beijing Institute of Technology, Jiaxing 314019, China
* Correspondence: 7520210145@bit.edu.cn (G.M.); xintang@bit.edu.cn (X.T.)
† These authors contributed equally to this work.

Abstract: Multi-color light-emitting diodes (LEDs) with various advantages of color tunability, self-luminescence, wide viewing angles, high color contrast, low power consumption, and flexibility provide a wide range of applications including full-color display, augmented reality/virtual reality technology, and wearable healthcare systems. In this review, we introduce three main types of multi-color LEDs: the organic LED, colloidal quantum dots (CQDs) LED, and CQD–organic hybrid LED. Various strategies for realizing multi-color LEDs are discussed including red, green, and blue sub-pixel side-by-side arrangement; vertically stacked LED unit configuration; and stacked emitter layers in a single LED. Finally, according to their status and challenges, we present an outlook of multi-color devices. We hope this review can inspire researchers and make a contribution to the further improvement of multi-color LED technology.

Keywords: multi-color LED; OLED; QLED; CQD–organic hybrid LED; full-color display

Citation: Ma, S.; Qi, Y.; Mu, G.; Chen, M.; Tang, X. Multi-Color Light-Emitting Diodes. *Coatings* **2023**, *13*, 182. https://doi.org/10.3390/coatings13010182

Academic Editor: Saulius Grigalevicius

Received: 8 December 2022
Revised: 8 January 2023
Accepted: 11 January 2023
Published: 13 January 2023

1. Introduction

Light-emitting diodes (LEDs) are semiconductor optoelectronic devices that emit light when electrons and holes recombine under applied voltage [1–4]. Multi-color LEDs with various advantages of color tunability, self-luminescence, wide viewing angles, high color contrast, and low power consumption are considered as the new-generation full-color displays used in televisions, laptops, mobile phones, and augmented reality/virtual reality technology [3]. In addition, the excellent flexibility and thin thickness of multi-color LEDs provide a wider range of applications in emerging optoelectronic devices including folding full-color displays, smartwatches, and wearable healthcare systems [5]. Furthermore, multi-color LED integration with the sensor is a promising and innovative way to visualize the electronic output signal of sensors to different visible colors, which has great application prospects in various applications such as real-time electrocardiogram monitors and electronic skin sensors [6,7].

Organic LEDs (OLEDs) are a type of LED in which the emissive electroluminescent layer is a film of the organic compound [7–12]. In 1950, French chemist André Bernanost for the first time discovered the electroluminescent properties of organic compounds; since then, tremendous progress has been made in the development of OLED technology [4,13,14]. Up to now, OLEDs have been mass-produced and commercialized, and successfully applied in either rigid or flexible OLED displays and excellent solid-state lighting [15,16]. Due to the mature application foundation of monochrome OLEDs for years, multi-color OLEDs with various lateral/vertical integration configurations have been fairly widely researched [17–20].

In recent years, colloidal quantum dots (CQDs) have been considered to be promising visible emitter materials to replace organic luminescent materials due to their inherent luminescent properties, including tunable emission wavelength, narrow spectral bandwidth, and high photoluminescence quantum yield [21–25]. Using CQDs as the emissive electroluminescent layer of LEDs, CQD-based LEDs (QLEDs) have stimulated a great deal of research interest. QLEDs possess unique merits of high color saturation, tunable emission color, high brightness, and simple solution processability [26–28]. Recent advances have enabled monochromatic QLEDs to exhibit excellent performance. For example, the external quantum efficiency (*EQE*) of red, green, and blue QLEDs is higher than 20%, 22%, and 18%, respectively [27,29–31]. On this basis, researchers began to aim for high-performance multi-color QLEDs to achieve high-resolution full-color QLED displays [32–35].

However, the unstable blue QLEDs with a much shorter lifetime (T_{50} lifetime of 200 h at an initial brightness of 1000 cd m^{-2}) than the red (T_{50} lifetime of 26,500 h at 1000 cd m^{-2}) and green ones (T_{50} lifetime of 25,000 h at 1000 cd m^{-2}) restrict the development of full-color QLEDs [27,29–31]. By substituting the blue QLEDs with relatively stable blue OLEDs, a CQD–organic hybrid multi-color LED has been proposed combining the excellent saturation of QLEDs and high stability of OLEDs [36,37].

There are many reviews about introducing monochrome OLEDs or QLEDs aiming to improve the device performance such as luminance, *EQE*, and lifetime. In early reviews, Vanessa Wood et al. summarized the features of various QLEDs whose structures were classified by different charge transport layers, outlined the challenges, and discussed the directions of QLED research [38]. Recently, Gloria Hong et al. made an overview of the history and development of the emitter materials of OLEDs including fluorescence, phosphorescence, and thermally activated delayed fluorescence, and explored the direction of new-generation luminescent materials [39]. From the perspective of solid-state lighting applications, Ramchandra Pode classified OLED devices based on the emission mechanism of emitting materials, and they paid attention to the low manufacturing costs and energy consumption of OLED technologies [40]. In addition, many reviews have been reported to research the luminous mechanism, summarize the device performance, and discuss the future development of monochrome OLEDs or QLEDs [41–54]. While reviews of monochrome LEDs are comprehensive, there is little focus on multi-color LEDs.

Many studies have reported various methods to realize multi-color LEDs, such as lateral/vertical LED integration configuration, and optimized multi-color emission layers. It is necessary to summarize and analyze the features of different methods, which are of great significance for the further improvement of multi-color LED technology. Recently, Bernard Geffroy et al. made an overview of full-color OLEDs from the perspective of different materials and discussed the development of multi-color OLED display technology [13]. Yizhe Sun et al. focused on the red/green/blue pixel patterning methods for realizing full-color QLED displays including inkjet printing, contact and transfer printing, and photolithography technologies [55]. In order to study the electroluminescence (EL) characteristics of multi-color QLEDs, the *EL* mechanisms, exciton formation process, and nonradiative processes were discussed in a review by Qilin Yuan et al., which has guiding significance to the optimal structure design [56]. In recent years, various types of multi-color LEDs, such as multi-color OLEDs, QLEDs, and CQD/organic hybrid LEDs, have appeared successively. However, to our knowledge, existing reviews about multi-color LEDs just focus on a single type of multi-color LED, such as multi-color OLEDs or QLEDs. There are few comparative summaries of various types of multi-color LEDs including multi-color OLEDs, QLEDs, and CQD/organic hybrid LEDs in one review. In addition, previous reviews have not systematically summarized and compared the different structures of multi-color LEDs, preventing a more comprehensive understanding of multi-color LED technology.

Thus, in this review, we introduce in detail three main types of multi-color LEDs: multi-color OLEDs, multi-color QLEDs, and multi-color CQD–organic hybrid LEDs. It is mainly divided into three representative device structures: (1) red, green, and blue sub-pixel side-by-side arrangement, (2) vertically stacked LED unit configuration, and (3) stacked emitter layers in a single LED. Progress in multi-color devices is summarized in Figure 1. Aside from this, the application of multi-color LEDs in flexible and wearable fields is briefly introduced. Finally, we present existing challenges and a vision for the future of multi-color devices. This review provides a more comprehensive summary, more detailed classification, and more diverse possible applications of multi-color LEDs compared with previous reviews, which is beneficial for researchers to contribute to the further improvement of multi-color technology.

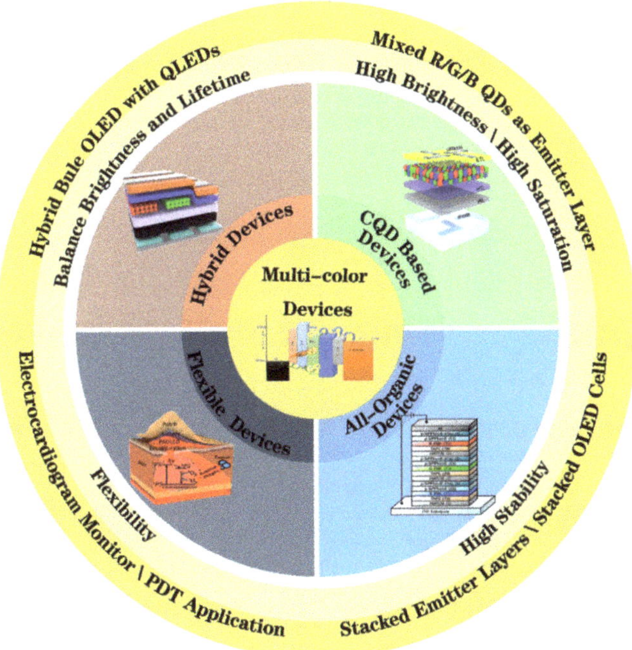

Figure 1. Progress in multi-color devices (PDT: photodynamic therapy). Reprinted with permission from Ref. [7]. 2020, ACS Nano. Reprinted with permission from Ref. [18]. 2018, ACS Photonics. Reprinted with permission from Ref. [33]. 2015, ACS Nano. Reprinted with permission from Ref. [36]. 2022, Springer Nature.

2. Types of Multi-Color Devices

After recent developments, there are many ways to implement multi-color devices, such as multi-colored OLEDs, QLEDs, and CQD–organic hybrid LEDs. In the following sections, we describe these types of multi-color LEDs, introduce representative works, and analyze their features. The summarization of multi-color devices is shown in Table 1.

2.1. Multi-Color OLEDs

OLEDs are expected to play a critical role in solid-state lighting and display application, owing to their unique properties, such as a wide range of colors, excellent luminance, ultrathin thickness, and flexibility. There are different ways to realize multi-color OLED devices. Firstly, the red, green, and blue OLEDs are arranged in parallel to achieve a wide color gamut [13,57]. However, this method has an intrinsic deficiency of a low pixel density. Secondly, the vertical stack OLED configuration has an improved pixel density and high efficiency. However, the structure of the device is relatively complex, leading to poor stability. Thirdly, the emission of OLEDs with multiple emission layers is selectively activated by different voltages [18,28,58–63]. Still, the transition of the complex exciton region between the two emission layers tends to reduce the device's efficiency.

2.1.1. Side-by-Side Structure

Combining the two, three, or four single-color OLED units in the side-by-side arrangement was proposed, driven by circuitry comprising a thin-film transistor (TFT) and capacitors that can address each pixel independently. The devices exhibit good optical performances since the light directly emits from respective units.

In 2003, C. David Muller et al. used a polymer with photoresist properties to make red/green/blue side-by-side OLED units via solution processing, as shown in Figure 2a [17]. The polymers are the classes of oxetane-functionalized spirobifluorene-co-fluorene, and they can be crosslinked photochemically to produce insoluble polymer networks in desired areas. A photograph of a red, green, and blue device is shown in Figure 2b. The obtained red-, green-, and blue-emitting units possess high luminous efficiency (*LE*) of about 1, 7, and 3 cd A^{-1}, respectively (Figure 2c).

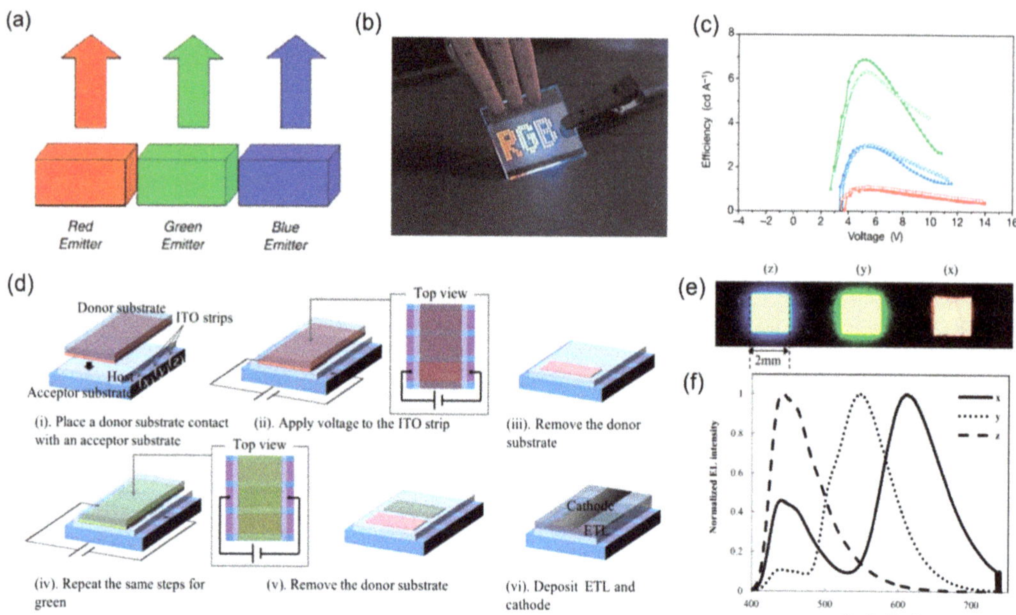

Figure 2. Side-by-Side Structure of Multi-color OLEDs. (**a**) Directly combine the red, green, and blue emitters in parallel. (**b**) Photograph of a red, green, and blue device. The dimensions of the glass substrate are 25 × 25 mm. (**c**) The *LE* for red-emitting, green-emitting, and blue-emitting devices. Reprinted with permission from Ref. [17]. 2003, Springer Nature. (**d**) A schematic diagram illustrating the steps of the procedure followed for fabricating red, green, and blue OLEDs. (**e**) *EL* images and (**f**) *EL* spectra of the devices. Reprinted from [64]. 2015, Optics Express.

Red, green, and blue color patterning is achieved via the sequential vacuum deposition of red, green, and blue materials through fine metal mask technology. However, the technique has several inherent limitations, such as difficulties in mask-to-substrate overlay alignment and in making micrometer-level dimensional accuracy masks, resulting in a low manufacturing yield and poor display resolution. In 2015, Yoshitaka Kajiyama et al. developed a maskless red/green/blue color patterning technique based on the diffusion of luminescent dopant molecules, which overcomes challenging issues in multi-color OLED display manufacturing arising from shadow mask limitations (Figure 2d) [64]. The *EL* images and corresponding *EL* spectra of the successfully fabricated red, green, and blue OLEDs side by side on one substrate are displayed in Figure 2e,f, respectively.

2.1.2. Vertically Stacked OLEDs

Compared with side-by-side red/green/blue OLED units to make a multi-color display, vertically stacked OLED units enable smaller pixel size and more fill factor, resulting in a threefold increased pixel density of the display. However, because of the light loss and the resistance increase in the semitransparent central electrode, the brightness of multi-color OLEDs inevitably decreases and a high drive voltage is needed.

Two-Unit Stacked OLEDs

In 1997, P.E. Burrows and S.R. Forrest et al. at Princeton University first proposed a new type of display pixel in which the two OLED units of red, green, or blue emission were placed in a vertically stacked geometry to mix red, green, or blue colors (Figure 3a) [65]. For example, independently blue and red emission OLED units are stacked and connected by the central magnesium–silver–indium tin oxide (Mg-Ag-ITO) electrode. The central electrode works as a common carrier injection layer of two units, and the color continuously changes from deep red to blue depending on the voltage bias ratio. On this basis, in 2009, C.J. Liang et al. achieved highly efficient multi-color emission with two-unit stacked OLEDs (Figure 3b) [9]. They found that the thin organic film underlying the ITO layer would be harmed by the magnetron sputtering process; thus, they solely adopted an aluminum–gold (Al-Au) dual film with suitable thickness as the inter-mediate electrode and, to improve the balance of electrons and holes injection, at both ends of the electrode they increased lithium fluoride (LiF) as the electron injection buffer layer and F_4TCNQ-doped m-MTDATA as the hole injection buffer layer.

However, adopting a traditional metallic conductor as an intermediate layer does not only reduce transparency but also influences stability. An effective way to reduce the driving voltages of tandem structures is by replacing them with a charge generation layer (CGL). In 2013, Yongbiao Zhao et al. from Nanyang technological university inserted a p-type layer of molybdenum trioxide (MoO_3) between hole transport layer TCTA constituting an n-i-p-i-n heterojunction symmetric structure (Figure 3d) [66]. When applying an alternating current (AC) signal, the n-i-p junction and p-i-n junction would switch on alternatively, and the device-emitted light mixed with two stacked OLED units. The correlated color temperature (CCT) of this device could be changed from 7575 to 2773 K, and the Commission Internationale de l'Eclairage (CIE) was varied from (0.16, 0.32) to (0.61, 0.38). Although the efficiency of carrier injection has been promoted, the diffusion of dopants would deteriorate the properties of devices. In 2022, Qian Chang et al. used two purely organic materials HAT-CN and CuPc composing a p-n type CGL as shown in Figure 3c; this device not only had better current efficiency but also realized color-tunable stacked OLEDs in a single direct current (DC) signal. When applying a driving voltage, p-n type CGL would generate two type carriers, and two OLED units emitted light simultaneously; the *CE* and *EQE* of obtained QLEDs can reach up to 46.3 cd A^{-1} and 15.1%, respectively [67].

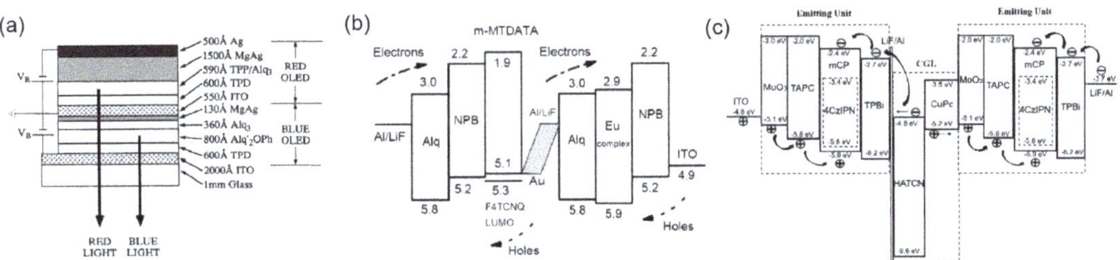

Figure 3. *Cont.*

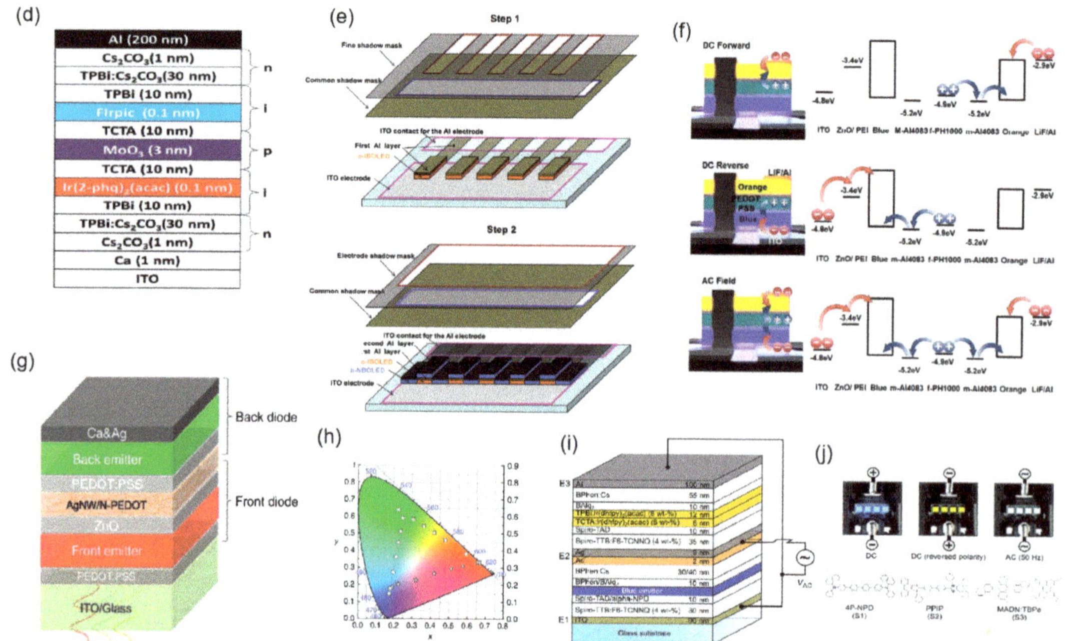

Figure 3. Two−Unit Stacked OLEDs. (**a**) Schematic cross−section of the layers of a red−blue tunable OLED. Reprinted with permission from Ref. [65]. 1996, American Institute of Physics. (**b**) The energy level diagram shows the highest occupied molecular orbital (HOMO) and the deeper lowest unoccupied molecular orbital (LUMO) levels, and carrier flows in OLEDs. Reprinted with permission from Ref. [9]. 2009, Elsevier. (**c**) The working principle of the tandem device. Reprinted with permission from Ref. [67]. 2022, Elsevier. (**d**) Schematic n-i-p-i-n structure of the OLEDs Reprinted with permission from Ref. [66]. 2013, Elsevier. (**e**) Fabrication steps of the proposed color−tunable OLED. The o−IBOLED is first deposited with a fine mask, followed by depositing the b−NBOLED without the fine mask. Reprinted with permission from Ref. [12]. 2013, Elsevier. (**f**) The operation mechanism of top orange and bottom blue tandem LEDs coupled with energy−level alignment under DC forward, DC reverse, and AC fields. Reprinted with permission from Ref. [68]. 2016, The Royal Society of Chemistry. (**g**) The schematic architecture of the color−tunable OLED device. Reprinted with permission from Ref. [19]. 2017, Springer Nature. (**h**) *CIE* coordinates of the three color-tunable devices calculated from their *EL* spectra of the tandem OLED cells with material combinations of green−blue, red−blue, and red−green operated by applying an AC signal with different pulse offsets. Reprinted with permission from Ref. [19]. 2017, Springer Nature. (**i**) Schematic illustration of the architecture of the investigated color−tunable AC/DC OLEDs. Reprinted with permission from Ref. [20]. 2015, Springer Nature. (**j**) Photographs showing an AC/DC OLED sample upon application of a DC bias (blue emission), a DC bias with reversed polarity (yellow emission), and an AC voltage with a frequency of 50 Hz (white emission). Reprinted with permission from Ref. [20]. 2015, Springer Nature.

For the practical application, there are issues regarding the complicated and expensive manufacturing process that need to be addressed. In 2013, Jiang et al. from Hong Kong University also fabricated a color-tunable device based on the two stacked OLED units without fine mask alignment, as shown in (Figure 3e) [12]. Although adjusting the polarity of the AC signal, the *CIE* coordinates are varied from (0.21, 0.23) to (0.57, 0.37). The color rendering index (*CRI*) is 54.2 when the *CIE* is (0.38, 0.29). The solution process for the polymer emitters and the interface materials can result in a much simpler and cheaper fabrication process. In 2016, Sung Hwan Cho et al. combined two tandem OLEDs by sharing a polymer electrode as a charge injection layer through solution process methods (Figure 3f) [68]. The maximum current efficiency (*CE*) and luminance of obtained OLEDs are 2.5 cd A^{-1} and 1300 cd m^{-2}, respectively. In 2017, Fei Guo et al. introduced highly conducting and transparent silver nanowires as the intermediate charge injection layer between two OLED units (Figure 3g) [19]. When a low-frequency square-shaped AC signal with positive and negative potentials is applied to the tandem cells, the two diodes can alternatingly turn on and emit pulses of the two primary colors. The *CIE* coordinates of the three color-tunable devices with material combinations of green–blue, red–blue, and red–green operated by applying an AC signal with different pulse offsets are shown in Figure 3h.

Another way to achieve multi-color emission with two OLED units is to introduce a yellow emission. In 2015, Markus Frobe et al. demonstrated a new device that could emit color from deep blue to warm white to saturate yellow by independently addressing a fluorescent blue emission unit and a phosphorescent yellow emission unit (Figure 3i,j) [20]. The leading way is the AC/direct current (DC) method by changing the pulse width and pulse height of the positive and negative cycles. The highest power efficiency (*PE*) values of OLED reach 36.8 lm W^{-1} at warm white color coordinates.

Three-Unit Stacked OLEDs

Three-unit stacked OLED devices are proposed due to the better color gamut and higher *CRI* compared with the two-unit stacked ones. However, the higher driving voltages for independently controlling three units are inevitably applied. Aside from this, serious color distortion issues appear caused by the micro-cavity effect.

In 2017, Mi Jin Park et al. successfully fabricated highly efficient three-stacked OLEDs based on the optical simulation results (Figure 4a); the measured spectrum was almost identical to the simulated device as shown in Figure 4b, showing an *EQE* of 49.4%, a *PE* of 33.4 lm W^{-1}, and a *CRI* of 93 (Figure 4c) [18]. In 2018, Hyunkoo Lee et al. also fabricated multi-color OLEDs with a vertical stack of three primary colors [61]. To reduce the driving voltage, they inverted the middle green OLED which shared metal electrodes with the bottom and the top OLEDs (Figure 4d). The driving voltages of blue, green, and red emission units are around 5.3 V, 2.8 V, and 4.4 V, respectively, at 1000 cd m^{-2} (Figure 4e). The *CCT* could be easily changed from warm white to cool white and a *CRI* of 88.7 could be reached. Furthermore, they also optimized the thickness of the hole transport layers (HTLs) and the electron transport layer (ETL) to alleviate the micro-cavity effect. As a result, the color distortion issues are alleviated and the *EQE* of blue, green, and red emission units can reach 11.1%, 10.9%, and 9.6%, respectively (Figure 4f).

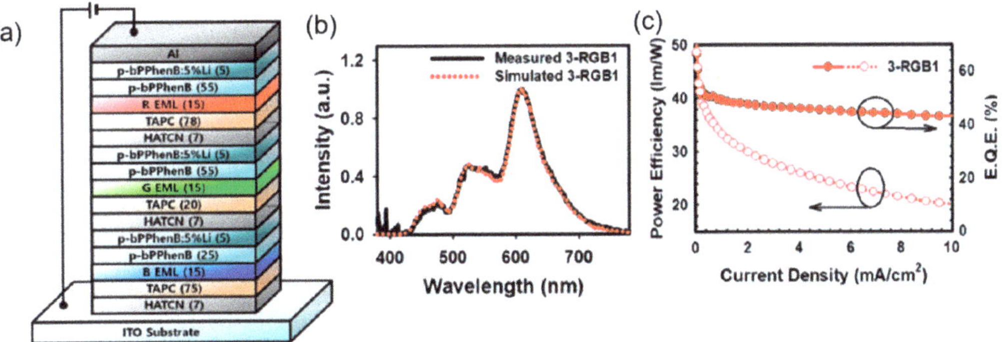

Figure 4. *Cont.*

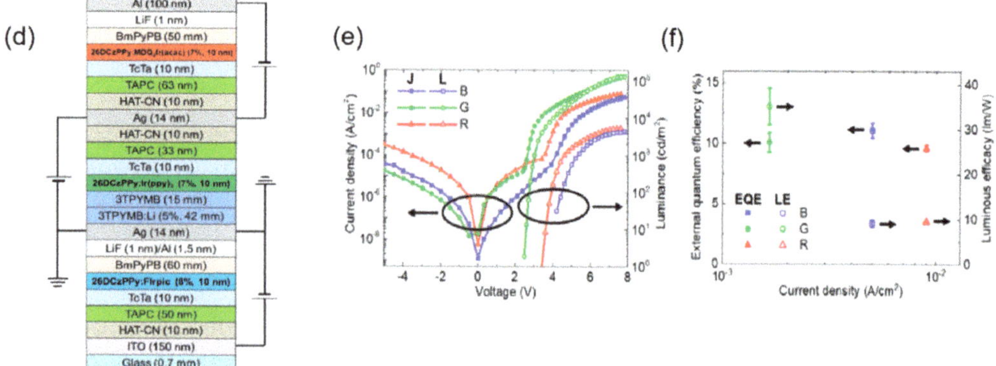

Figure 4. Three-Unit Stacked OLEDs. (**a**) Schematic structure of the color-tunable OLED. (**b**) Comparison of measured and simulated *EL* spectra of the device. (**c**) *PE* and *EQE* versus the current density of the device. Reprinted with permission from Ref. [18]. 2017, ACS Photonics. (**d**) Schematic structure of the color-tunable OLED with independently controlled red, green, and blue OLEDs. (**e**) Luminance and current density versus voltage characteristics of independently controlled red, green, and blue OLEDs in the color-tunable OLED. (**f**) *EQE* and LE of independently controlled red, green, and blue OLEDs in the color-tunable OLED. Reprinted from Ref. [61]. 2018, Optical Express.

2.1.3. Stacked Emitter Layers

Stacking various emitter layers with different emission wavelengths is a direct and simple way to achieve multi-color emission compared with side-by-side and vertically stacked OLED units. In this way, multiple colors can be obtained by changing the bias voltages. To make carriers' injection more balanced and emission of OLEDs more efficient with stacked emission layers, the energy level structure should be carefully designed and the energy transfer between the host and dopant should be considered.

Single Emission Layer

The multi-color OLED with a single emission layer (S-EML) has a simple structure, and it can be fabricated by an all-solution method suitable for commercial applications. By doping multi-emission wavelength materials into the host emission layer, the different colors can be emitted by changing the bias of voltage. In addition, the emission spectrum is influenced by the doped ratios; therefore, the concentration of dopants should be considered and optimized.

In 1994, J. Kido and K. Hongawa et al. first proposed the white OLEDs by using a single emitter layer doped with three fluorescent dyes [69]. The fluorescent dyes including blue-emitting 1,1,4,4-tetraphenyl-1,3-butadiene, green-emitting coumarin 6, and orange-emitting DCM 1 were doped into the poly(9-vinylcarbazole) (PVK) layer (Figure 5a). The obtained OLEDs possessed high luminance of 3400 cd m^{-2} at a driving voltage of 14 V and an excellent LE of 0.85 lm W^{-1}.

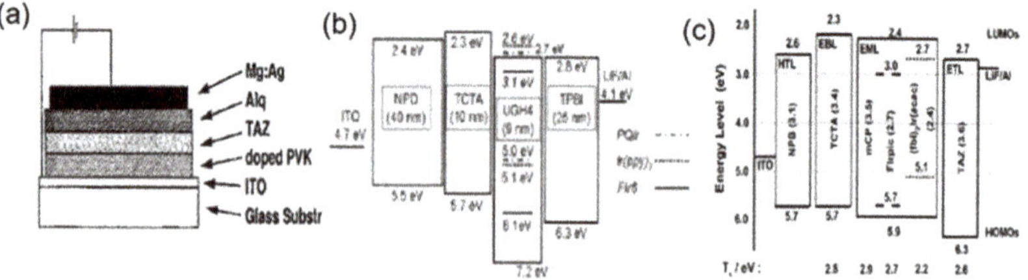

Figure 5. *Cont.*

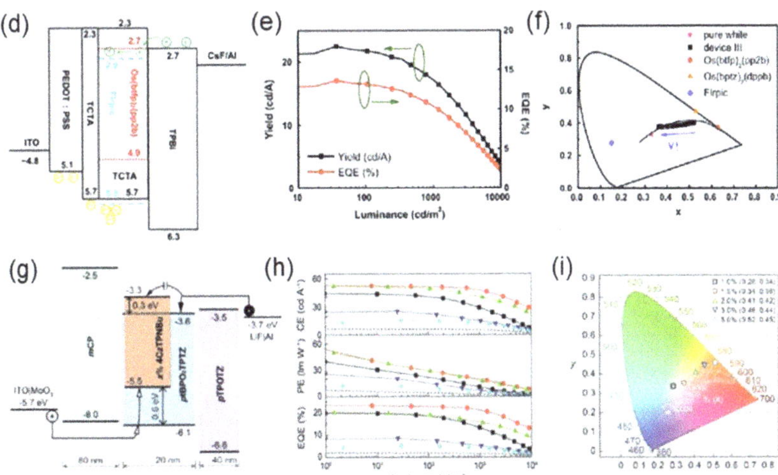

Figure 5. Single Emission Layer of Multi-color OLEDs. (**a**) The configuration of the OLED cell. Reprinted with permission from Ref. [69]. 1994, Applied Physics Letters. (**b**) The schematic diagram of the energy structure. Reprinted with permission from Ref. [70]. 2004, Wiley-VCH. (**c**) Proposed energy diagram of the white OLED. Reprinted with permission from Ref. [71]. 2008, John Wiley and Sons. (**d**) Energy band diagram of the materials in the device. (**e**) *CE* and *EQE* characteristics versus luminance of device. (**f**) *CIE* 1931 coordinates shifting with applied voltages from 4.7 V to 5.0 V. Reprinted with permission from Ref. [72]. 2015, Elsevier (**g**) The schematic diagram of the energy structure. (**h**) *CE*, *PE*, and *EQE* versus luminance of OLEDs with different blue TADF doping concentrations. (**i**) *CIE* 1931 coordinates' dependence of the devices on the concentration of dopant. The black body locus and the color temperature lines were added for reference. Reprinted with permission from Ref. [73]. 2020, John Wiley and Sons.

Compared with fluorescent materials, OLEDs with phosphorescent materials as emission layers possess higher efficiency. In 2004, Brian W et al. doped phosphorescent materials of red-emitting PQIr, green-emitting $Ir(ppy)_3$, and blue-emitting FIr_6 into the inert host UGH2 (Figure 5b) [70]. The *CIE* chromaticity coordinates of obtained multi-color OLEDs vary from (0.43, 0.45) at 0.1 mA cm^{-2} to (0.38, 0.45) at 10 mA cm^{-2}. In addition, the maximum *PE* could be up to 42 lm W^{-1}, and the *CRI* is as high as 80. In 2009, Qi Wang et al. used FIrpic as the blue-emission doping phosphorescent material and $(fbi)_2Ir(acac)$ as an orange one [71]. The redundant excitons are formed on the host of mCP, and then the energy is transferred to FIrpic via the Forster- or Dexter-type transfer mechanism, leading to the blue emission (Figure 5c). Thus, the charge balance is improved and the common energy losses are eliminated. The performance of OLED is enhanced with *PE* of 42.5 lm W^{-1}, *EQE* of 19.3%, and *CE* of 52.8 cd A^{-1}.

Although the performance of OLEDs with phosphorescent materials as emission layers has been greatly improved, the rare metal in phosphorescent materials increases the cost of preparation. Recently, OLEDs based on thermally activated delayed fluorescence (TADF) have received wide attention due to both high efficiency and low cost. In 2015, Huang et al. fabricated color-tunable OLEDs through the solution-processed method with TCAC as the host material, FIrpic as the blue dopant, Os(btfp)2(pp2b) as the red dopant, and Os(bptz)2(dppb) as the yellow dopant of light-emitting layers (EMLs) (Figure 5d) [72]. The *EQE* of multi-OLEDs is up to 13.6%, as shown in Figure 5e. In addition, the OLEDs exhibit variation in the color of *CIE* from (0.52,0.40) to (0.36,0.38) (Figure 5f). In 2017, Jun-Yi Wu et al. reported a T-P (TADF-phosphorescence hybrid white OLED using the TADF material of DMAC-TRZ as blue dopants, and the phosphorescent materials of Ir(dpm)PQ2 as red dopants [62]. The obtained OLEDs possess excellent performance of *EQE* of 12.32% and

PE of 18.1 lm W^{-1}. To further improve the efficiency of the devices, in 2020, Ding D et al. demonstrated an all-TADF-doped multi-color OLED [73]. The EML consists of yellow TADF dopants (4CzTPNBu) and blue TADF ones (ptBCzPO2TPTZ). The low-energy-gap yellow dopants with the shallower HOMO and LUMO could form the hole and electron traps for carrier and exciton capture, therefore further hampering the charge exchange-based exciton migration (Figure 5g). The devices possess high *PE* of 55.1 lm W^{-1}, excellent *EQE* of 23.6%, and maximum luminance beyond 30000 cd m^{-2}, as shown in Figure 5h. At 1000 cd m^{-2}, the OLEDs with different blue TADF doping concentrations of 1.0%, 1.5%, 2.0%, and 3.0% emit cool white, pure white, and warm white lights with the *CIE* coordinates of (0.28, 0.34), (0.34, 0.36), (0.41, 0.42), and (0.46, 0.44) and the *CCT* of 8332, 5152, 3563, and 2883 K, respectively (Figure 5i).

Double Emission Layers

Multi-color OLEDs with a double emission layer (D-EML) concept were first introduced by X. Zhou et al. in 2002 [74]. For OLEDs with S-EMLs, electrons or holes tend to accumulate at the interface of the EML/ETL or EML/HTL due to the large energy barrier. The high density of accumulated carriers leads to the quenching of triplet excitons. In contrast, the D-EML structure can significantly avoid carrier accumulation at the interface by widening the triplet excitons' generation zone, leading to better device performance.

In 2004, Gufeng He et al. fabricated an efficient OLED with D-EML structure through doping the same phosphorescent material $Ir(ppy)_3$ into both TCTA and TAZ emission hosts (Figure 6a) [75]. Because the HOMO level of the TCTA was much lower than that of $Ir(ppy)_3$, a part of the holes would be captured by the $Ir(ppy)_3$; in this way, excitons not only formed on the interface of TCTA and TAZ, but also could be combined in the $Ir(ppy)_3$ sites directly. This structure increased the recombination area, resulting in an excellent performance with *PE* of 64 lm W^{-1} at 1000 cd m^{-2} and *EQE* of 19.3% (Figure 6b). To broaden the range of colors, in 2010, Sebastian Reineke et al. embedded the blue and green dopants in a common host of TPBi, and the red one in the host of TCTA to achieve D-EML-structured multi-color OLEDs (Figure 6c) [76]. The *PE* of obtained OLEDs can reach 90 lm W^{-1} at the brightness of 1000 cd m^{-2}, and the *EQE* can reach 34% (Figure 6d). In 2016, Xuming Zhuang et al. reported a four-color OLED device employing a D-EML structure with a blue host/orange dopant and green host/red dopant; they avoided using the structure of only the dopant as the emitting molecules, and achieved the most broad range of spectra with the D-EML structure, exhibiting a high *CRI* of 92, *EQE* of 23.3%, and *PE* of 63.2 lm W^{-1} (Figure 6e,f) [59].

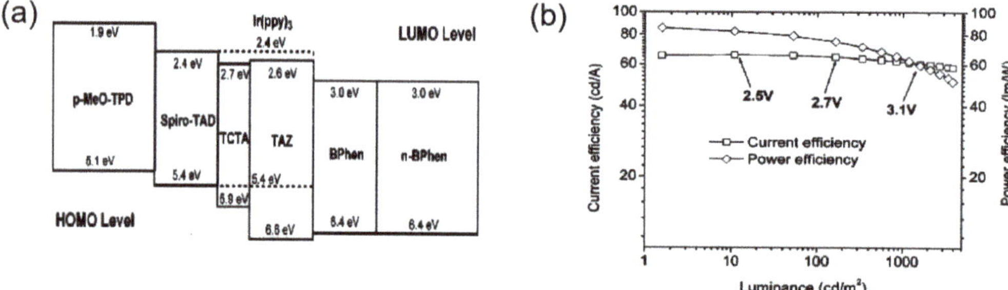

Figure 6. *Cont.*

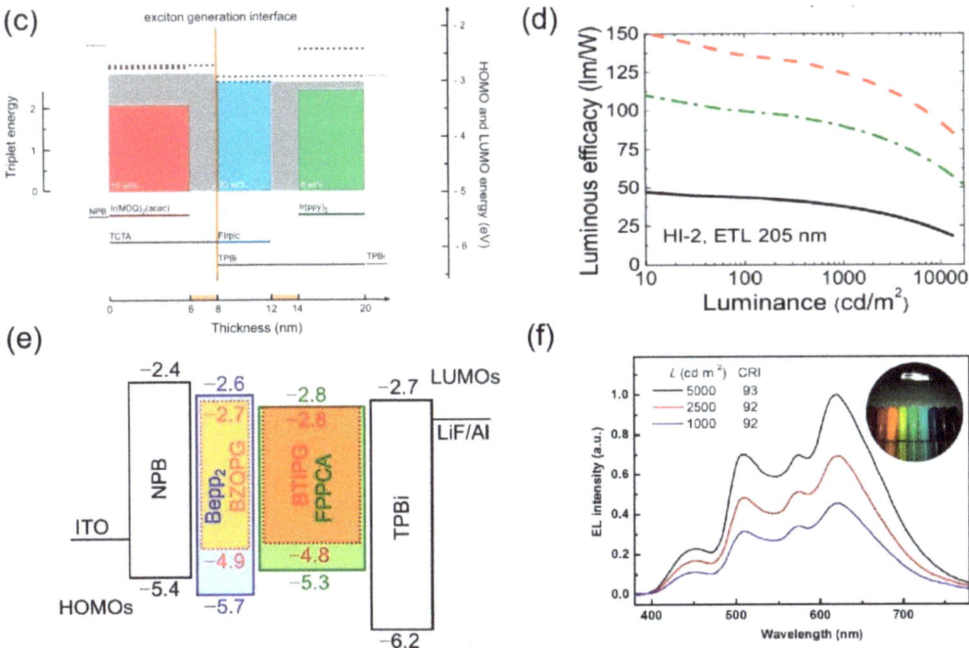

Figure 6. Double Emission Layers of Multi-color OLEDs. (**a**) Device structure of D-EML p-i-n OLED and the proposed energy level diagram. (**b**) The current efficiencies (square) and power efficiencies (diamond) of an optimized D-EML OLED. Reprinted with permission from Ref. [75]. 2004, AIP Publishing. (**c**) Energy diagram of the materials used in OLEDs. (**d**) *EL* spectra and the corresponding *CRI* values of the device at different luminance. Reprinted with permission from Ref. [76]. 2009, Springer Nature. (**e**) Emission layer energy level diagram. HOMO and LUMO levels are plotted as solid and dashed lines, respectively, and filled boxes refer to the materials' triplet energies. The orange line indicates the exciton generation interface. The orange x-axis marks intrinsic interlayers (**f**) The LE of the device as a function of luminance. These values are measured in three configurations: flat (without outcoupling solid), with an attached half-sphere (dash), and with an attached pyramidal pattern (dash–dot). Reprinted with permission from Ref. [59]. 2016, ACS.

Multiple Emission Layers

To further improve the balance of the carrier injection, it is necessary to develop color-tunable OLEDs with multiple emission layers. In 2016, Ping Chen et al. from Jilin University fabricated OLEDs with double blue emission layers and an orange ultrathin layer between them [60]. TCTA with high hole transport mobility and TPBi with high electron transport mobility were blended as the mixed host for the blue phosphorescent emitter (Figure 7a). Two different hosts expand the exciton recombination zone, which leads to charge balance. Normalized *EL* spectra of the device at different voltages from 3 V to 9 V are shown in Figure 7b. The peak *PE* can be above 40.8 lm W^{-1}, and a *CRI* of 62 and *CE* of 39.8 cd A^{-1} can be achieved. In 2018, Gyeong Woo Kim et al. also adopted the blue–yellow–blue multiple EML structure while they selected the TADF material as a blue emitter (Figure 7c), achieving a high *EQE* of 23.1% at the *CIE* coordinate of (0.324, 0.337) [77]. Normalized *EL* spectra of devices are presented in Figure 7d. Although sandwiching single-color emission layers between two blue emission layers has obtained stable multi-color emission, it is essential to increase three or more emitters to achieve wider luminescent spectra. In 2019, Baiqian Wang et al. fabricated a relatively stable white OLED with two emission layers of both blue dopants, and three ultrathin layers of red, orange, and green emission layers in the middle (Figure 7e) [63]. The obtained OLEDs have ambipolar charge carrier transport properties between EMLs. Normalized *EL* spectra of the device at different driving voltages are shown in Figure 7f. The performance of OLEDs is excellent with *CRI* reaching 94 and a maximum *PE* of 33.4 lm W^{-1}.

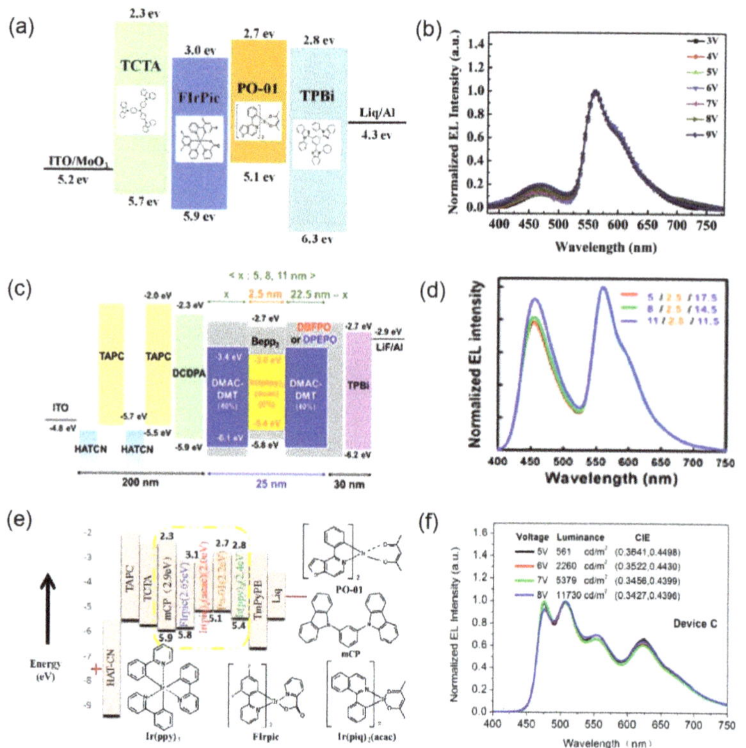

Figure 7. Multiple Emission Layers of Multi-color OLEDs. (**a**) The detailed energy level diagram and chemical structures of the materials. (**b**) Normalized *EL* spectra of the device at different voltages from 3 V to 9 V. Reprinted with permission from Ref. [60]. 2016, Elsevier. (**c**) Energy level diagram with the detailed device structure of OLEDs. (**d**) Normalized *EL* spectra of white devices. Reprinted with permission from Ref. [77]. 2018, Springer Nature. (**e**) The schematic diagram of devices and the energy level/molecular structure of part materials. (**f**) Normalized *EL* spectra of the device at different driving voltages. Reprinted with permission from Ref. [63]. 2019, Elsevier.

2.2. Multi-Color QLEDs

CQDs have been considered promising visible emitter materials to replace organic luminescent materials due to their inherent luminescent properties, including tunable emission wavelengths, narrow spectral bandwidths, and solution–process compatibility [24,25,27]. Since the first demonstration of CQD-based LEDs (QLEDs) in 1994, the performance of monochromatic QLEDs has been significantly improved [78]. Based on this, recently, researchers began to aim for high-performance multi-color QLEDs to obtain high-resolution full-color QLED displays.

2.2.1. Side-by-Side Structure

The realization of multi-color QLEDs is early achieved by the side-by-side patterning of red/green/blue CQDs onto the pixelated display panel. The cross-contamination issues of the red/green/blue pixels could be solved through the photolithography approach, inkjets method, or transfer printing process replacing the spin-coating method used to fabricate monochrome displays. However, the fill factors and pixel density are inevitably sacrificed due to the lateral integration configuration. Aside from this, other key figures of merit such as operation lifetime, film uniformity, and interfacial charge transport efficiency are probably deteriorated because of incompatibility with lithographical chemicals or coffee ring effects.

In 2011, Tae-Ho Kim et al. demonstrated large-area, full-color CQD displays driven by oxide-based thin-film transistor (TFT) arrays through side-by-side nano-transferring red/green/blue CQDs (Figure 8a) [79]. Solvent-free transfer printing associated with kinetic control and interfacial chemistry enables printed CQD films to exhibit excellent morphology, well-ordered CQD structure, and clearly defined interfaces. The fluorescence micrograph of the transfer-printed red/green/blue CQD stripes onto the glass substrate excited by 365 nm ultraviolet (UV) radiation is shown in Figure 8b. A 4-inch full-color active matrix CQD display with a resolution of 320 × 240 pixels can be realized (Figure 8c). Figure 8d shows an optical image of simultaneous red/green/blue *EL* emission from the pixelated area of QDs during operation.

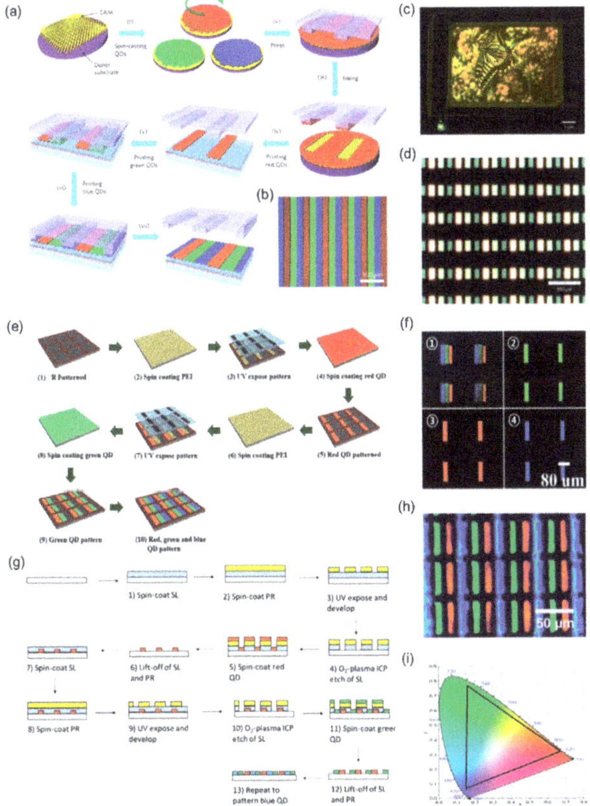

Figure 8. Side-by-Side Structure of Multi-color QLEDs. (**a**) Schematic of the transfer printing process for the patterning of CQDs. (**b**) Fluorescence micrograph of the transfer-printed red/green/blue CQD stripes onto the glass substrate, excited by 365 nm UV radiation. (**c**) *EL* image of a 4-inch full-color CQD display using a TFT backplane with a 320 × 240 pixel array. (**d**) Optical image of simultaneous red/green/blue *EL* emission from all pixels under operation. Reprinted with permission from Ref. [79]. 2011, Springer Nature. (**e**) Illustration of photolithography process for the patterning of CQDs. (**f**) PL image of patterned red/green/blue CQDs coated on different substrates (②,③, and④) and one substrate (①). Reprinted with permission from Ref. [32]. 2018, John Wiley and Sons. (**g**) Schematic illustration of patterning CQDs with different colors on a substrate via the photolithography approach. (**h**) *EL* image of the 500-ppi, full-color QLED. (**i**) Color coordinates of red/green/blue subpixels when lighted separately. Reprinted with permission from Ref. [34]. 2020, Springer Nature.

However, nonuniformity over the deposited area of the printing process still exists. The development of new fabrication technologies for efficient high-resolution and large-area patterning of CQD devices is required. In 2018, Han-Lim Kang et al. proposed efficient and simple patterning technologies that employ locally controlled surface tailoring of constitutional functional layers via photochemical deactivation routes (Figure 8e) [32]. The patterning area of each CQD layer was observed in the photoluminescence (*PL*) images as a continuous and rectangular region (20 μm × 80 μm) when excited by 356 nm UV radiation (Figure 8f). The obtained multi-color QLED devices possess a maximum luminescence of 1950 cd m^{-2} and a *CE* of 2.9 cd A^{-1}.

Although the technologies for high-resolution and large-area patterning of CQD devices have been improved, the relatively high cost limits the practical application. An optimized low-cost photolithography process seems to be a suitable practical patterning approach. In 2020, Wenhai Mei et al. demonstrated a sacrificial layer-assisted patterning (SLAP) approach, which could be applied in conjunction with photolithography to fabricate high-resolution, full-color side-by-side CQD patterns (Figure 8g) [34]. A 500-ppi, full-color, passive, matrix QLED prototype with no color impurities in the subpixels was successfully fabricated via this process (Figure 8h). The obtained QLED has a high color gamut of 114% (National Television Standards Committee (NTSC)) (Figure 8i).

2.2.2. Vertically Stacked QLEDs

Instead of directly patterning EMLs, the combination of white QLEDs and color filters can also realize red/green/blue side-by-side color pixels to avoid the problem of chemical incompatibility caused by photolithography and the printing process. Vertically stacking red, green, and blue QLED units to form a tandem structure offers a practical solution for obtaining white QLEDs with high *CE* and a long lifetime, combining with color filters to realize high-resolution full-color displays. However, the luminance of QLEDs is markedly weakened because over 2/3 of the emitted light is absorbed by color filters.

In 2017, Heng Zhang et al. demonstrated all-solution-processed three-unit (red/green/blue) white tandem QLEDs for the first time (Figure 9a) [80]. The tandem devices are achieved by serially connecting the red bottom sub-QLED, the green middle sub-QLED, and the blue top sub-QLED using the inter-connecting layer based on the $Zn_{0.9}Mg_{0.1}O$/poly(ethylenedioxythiophene): polystyrenesulfonate (ZnMgO/PEDOT: PSS) heterojunction (Figure 9b). The three-unit white QLEDs exhibit evenly separated red, green, and blue emissions and red/green/blue colors can be easily recovered using conventional color filters. A peak *CE* of 4.75 cd A^{-1}, *EQE* of 2.0%, and a high luminance of 4206 cd m^{-2} are obtained, as shown in Figure 9c.

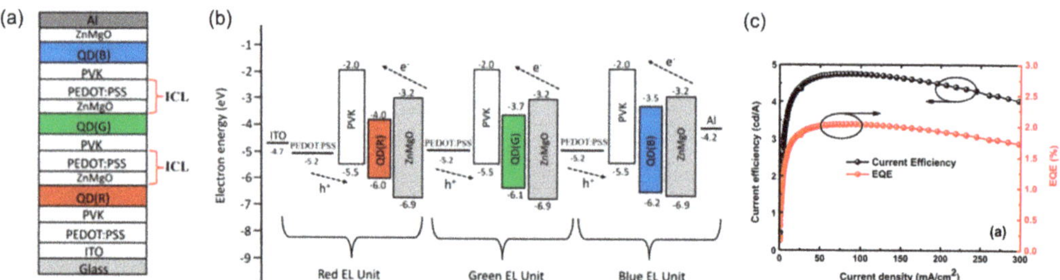

Figure 9. Vertically Stacked QLEDs. (**a**) The structure of three-unit tandem QLEDs. Copyright 2017, Society for Information Display. (**b**) Energy band diagram of the three-unit tandem QLEDs. (**c**) *CE*/*EQE* versus current density characteristics of the three-unit white tandem QLEDs. Reprinted with permission from Ref. [80]. 2017, John Wiley and Sons.

2.2.3. Stacked Emitter Layers

Developing color-tunable QLEDs with stacked emitter layers could circumvent the limitation of red/green/blue side-by-side color pixels. Because the full color is attained in a single color-tunable pixel instead of three red/green/blue pixels, the pixel density and fill factor of a display with color-tunable pixels can be enhanced three times. This is essential to improve the resolution of full-color QLED displays. Earlier studies began by vertically stacking multiple emission layers to achieve multi-color QLEDs. However, the fabrication of QLEDs with multiple emission layers is complex, which is difficult for repeatable and large-scale preparation. Some researchers have found that mixed CQDs as a single emission layer can also achieve highly efficient multi-color emission controlled by bias, which is closer to the practical application.

Double or Multiple Emission Layers

Patterning and stacking variously colored CQDs (red, orange, green, blue) in the exciton recombination zone could be an efficient method to acquire multi-colored QLEDs. However, for the realization of full-color displays with QLEDs, the development of a fabrication process for the deposition of homogeneous and uniform CQD layers over a large area with patterning capability is still necessary. In 2010, Wan Ki Bae et al. fabricated the all-CQD multi-layer films via a layer-by-layer assembly method using electrostatic interactions between each layer through the sequential deposition of oppositely charged CQDs onto the substrates (Figure 10a) [81]. The exciton recombination zone was investigated by monitoring the *EL* spectral change with the introduction of sensing layers within the all-CQD multi-layer films. The total *EL* emission comes mostly at the top CQD monolayer, adjacent to the ETL layers (~90%), and partially at the second CQD monolayer from the top (Figure 10b).

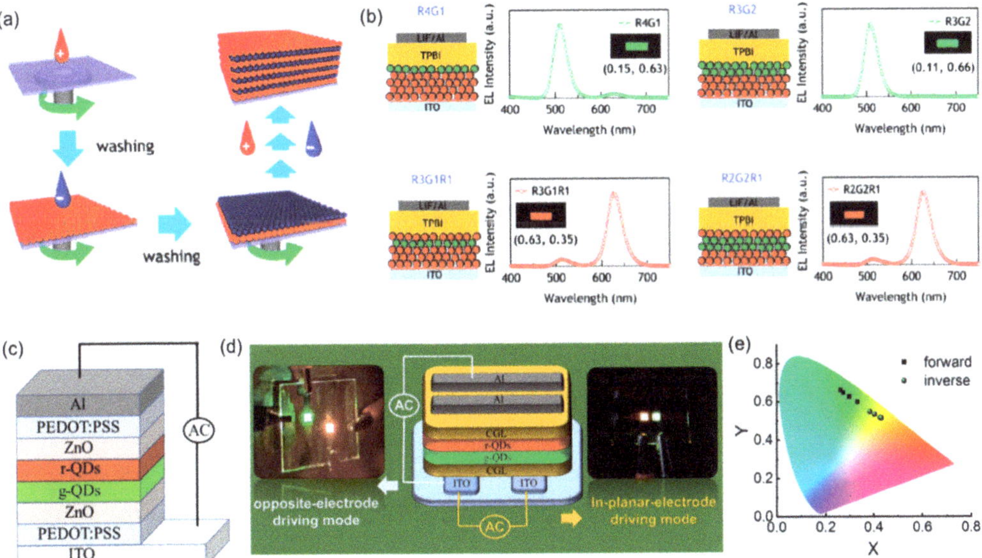

Figure 10. Double or Multiple Emission Layers of Multi-color QLEDs. (**a**) Schematic for the preparation of all-CQD multi-layer films based on spin-assisted layer-by-layer assembly by sequentially depositing oppositely charged CQDs. Copyright 2010 American Chemical Society. (**b**) Device structures and corresponding *EL* spectra of QLEDs possessing sensing CQD layers (green CQD layers) at various positions within the red CQD multi-layer films. The insets show images of each QLED (pixel size of 3.6 × 1.4 mm^2) and corresponding *CIE* indices of the *EL* spectra. Reprinted with permission from Ref. [81]. 2010, ACS. (**c**) Schematic structure of the all-solution-processed AC device. (**d**) Photos of a working AC QLED and corresponding driving diagrams in the opposite-electrode and in-planar-electrode driving manners. (**e**) *CIE* color coordinates of the AC device under different driving voltages and polarities. Reprinted with permission from Ref. [82]. 2021, ACS.

Although large-area practical multi-colored QLEDs are realized, the accumulated charges at the interfaces are always inevitable in the QLED devices. In 2021, Ting Wang et al. found that the charge accumulation issue could be addressed by employing the alternating current (AC) driving mode in a QLED, and they successfully fabricated multi-color QLEDs with bilayer stacked emissive layers composed of red and green CQDs (Figure 10c) [82]. The emission color of the AC QLED can be tuned by both the polarity and amplitude of the driving voltage (Figure 10d,e). The efficient electron/hole injection to the emission helps to form excitons and reduces the turn-on voltage.

Single Emission Layer with mixed CQDs

Although multi-color light emission is achieved by stacking double or multiple emission layers, the alternatingly consecutive solvent treatments inevitably damage the prior-deposited CQD films and degrade interfacial carrier transport efficiency. By using a single emission layer with mixed red/green/blue CQDs as light-emitting materials, highly efficient multi-color QLEDs that can emit full colors under different driving bias voltages are achieved.

In 2015, Ki-Heon Lee et al. reported all-solution-processed bi- and trichromatic QLEDs, where red, green, or blue CQD-mixed EMLs are sandwiched with poly(9-vinlycarbazole) (PVK, HTL) and zinc oxide nanoparticles (ZnO, ETL) with a standard device architecture (Figure 11a,b) [33]. *PL* decay dynamics of a homogeneously distributed red/green/blue CQD-mixed solid film were first examined, showing the shortened lifetimes for blue and green QDs and the lengthened lifetime for red ones as a result of Forster resonant energy transfer (FRET) among different band gap CQD emitters (Figure 11c). On this basis, in 2022, Ge Mu et al. achieved color-tunable QLEDs with the largest color variation controlled by bias among existing multi-color QLEDs by optimizing the mixing ratio of red, green, and blue CQDs [35]. Normalized *PL* spectra of mixed red/green/blue CQDs used for the fabrication of full-color tunable QLEDs with different formulations are shown in Figure 11d–f. At the optimal mixing ratio of red, green, and blue CQDs, full-color tunable QLEDs exhibit wide color variation ranging from the *CIE* chromaticity coordinates of red (0.649, 0.330) to orange (0.453, 0.389) to yellow (0.350, 0.347) to green (0.283, 0.305) to blue (0.255, 0.264) upon increasing voltages from 2 V to 9 V (Figure 11g–i). In addition, the fabricated multi-color QLEDs show high luminance approaching 10^3 cd m^{-2} and a superior *EQE* of 13.3%.

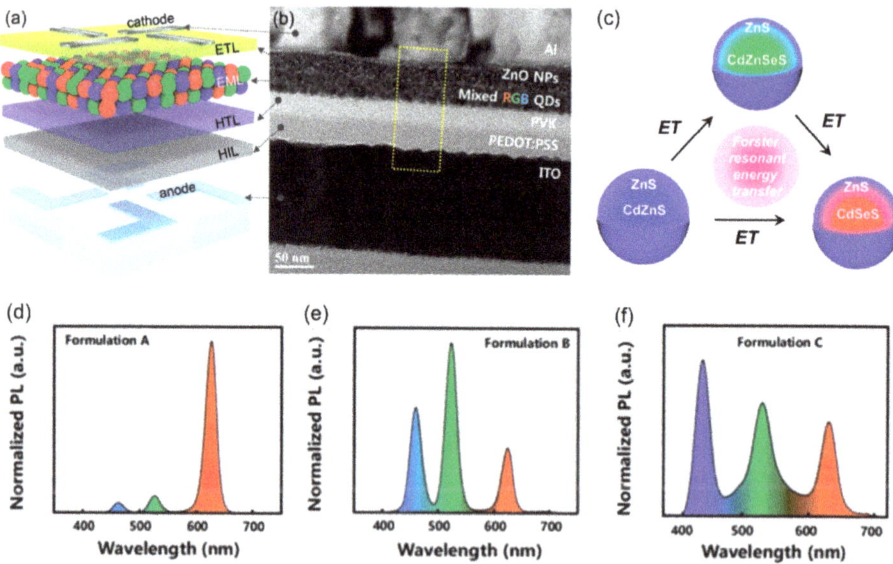

Figure 11. *Cont.*

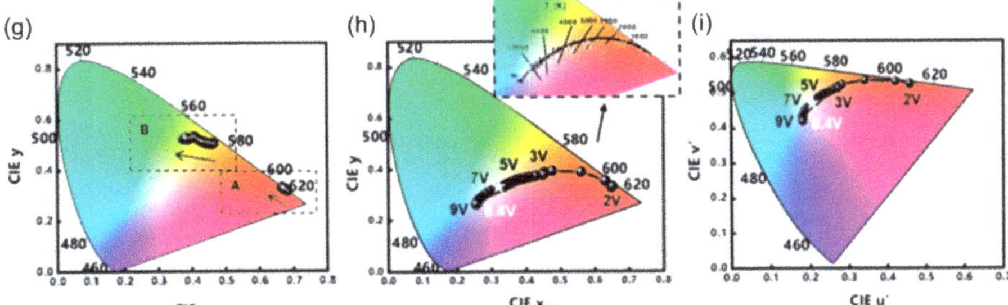

Figure 11. Single Emission Layer with mixed CQDs. (**a**) Device structure and (**b**) cross-sectional transmission electron microscopy (TEM) micrograph of all-solution-processed, multi-layered, full-color QLED. Copyright 2015 ACS Nano. (**c**) Schematic of inter-CQD FRET proposed based on PL decay dynamics. Reprinted with permission from Ref. [33]. 2015, ACS Nano. Normalized PL spectra of mixed red/green/blue CQDs used for the fabrication of full-color tunable QLEDs with (**d**) formulation A, (**e**) formulation B, and (**f**) formulation C. Evolution of *CIE* 1931 color coordinates of (**g**) formulation A and B, and (**h**) formulation of C-based full-color tunable QLEDs with increasing bias voltages. (**i**) Evolution of *CIE* 1976 color coordinates of formulation C-based full-color tunable QLEDs with increasing bias voltages. Reprinted from Ref. [35]. 2022 Chinese Laser Press.

2.3. *CQD and Organic Hybrid LED*

Although several methods are proposed to obtain high-performance multi-color QLEDs, the realization of high-resolution full-color QLED displays remains challenging. This is because the blue QLEDs are unstable, with a short T_{50} lifetime of 200 h at an initial brightness of 1000 cd m^{-2}, which is much shorter than those of red and green QLEDs of 26,500 and 25,000 h, respectively. However, blue OLEDs are relatively stable and have been applied in displays for years. By substituting the blue QLEDs with blue OLEDs, a hybrid device promises to achieve a multi-color LED with high saturation and high stability, which is beneficial for realizing high-resolution full-color displays.

In 2020, Heng Zhang et al. stacked a yellow QLED with a blue OLED using a transparent, indium–zinc oxide (IZO), intermediate connecting electrode to achieve a full-color tunable hybrid tandem LED (Figure 12a) [37]. By varying the driving AC signals, the device could emit red, green, and blue primary colors as well as arbitrary colors covering a 63% National Television System Committee (NTSC) color triangle (Figure 12b). Both the *EQE* of the blue OLED and the yellow QLED are high at 5.6% and 11.2%, respectively, and the maximum brightness of the blue OLED and yellow QLED is 9359 and 51,590 cd m^{-2}, respectively, showing that the introduction of IZO has no bad effect on the device performance (Figure 12c,d). The brightness/efficiency of the red and green emissions according to correlating the emission color with the voltage is shown in Figure 12e.

The tandem LED structure is relatively complex; thus, in 2022, Suhyeon Lee et al. developed a simpler method to fabricate a CQD–organic hybrid device [36]. They deposited an organic blue common layer (BCL) through a common mask over the green and red QLEDs (Figure 12f). The commonly deposited BCL is not only a blue EML but also an ETL for the red and green QLED sub-pixels (Figure 12g). Although the blue EML was shared, the *CIE* color coordinates of the red and green QLEDs were negligibly changed and thus we can keep the advantages of excellent color purity of CQDs (Figure 12h).

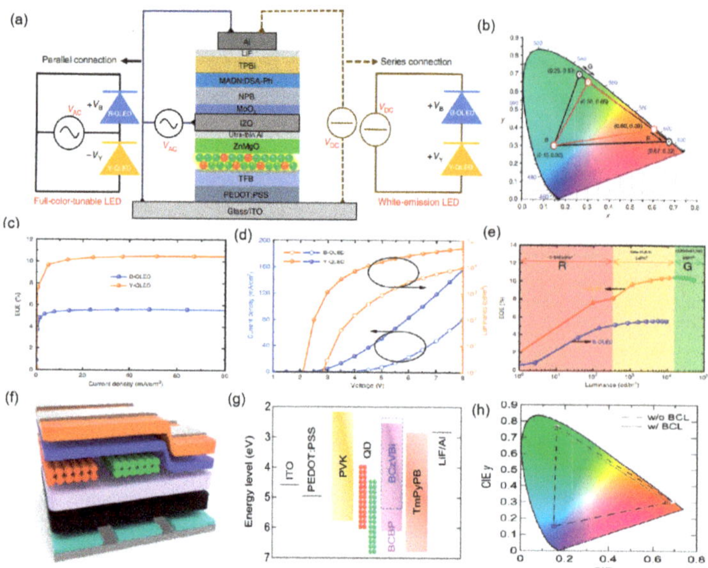

Figure 12. CQD and Organic Hybrid LED. (**a**) Schematic of the multifunctional tandem LED. (**b**) *CIE* chart and color coordinates of the red, green, and blue primary colors. (**c**) The *EQE* versus current density, (**d**) the current density and luminance versus voltage, and (**e**) the *EQE* versus luminance characteristics of blue OLED and yellow QLED, measured by extracting the IZO as a common electrode. Reprinted with permission from Ref. [37]. 2020, Springer Nature. (**f**) A schematic illustration of the hybrid full-color LED with red- and green-emitting QLED sub-pixels and a blue-emitting OLED sub-pixel (**g**) The energy level diagram of used materials. (**h**) *CIE* coordinates and color space. The solid and dashed lines connect the *CIE* coordinates of the devices with and without the BCL, respectively. Reprinted with permission from Ref. [36]. 2022, Springer Nature.

Table 1. Summarization of Multi-color Devices.

Year		**Type**	**Maximum Brightness (cd m^{-2})**	***EQE* (%)**	***PE* (lm W^{-1})**	***CE* (cd A^{-1})**	***CIE***	**Ref.**
2003	OLED	Side-by-Side Structure	-	-	-	7 2.9	Green blue	[17]
2014	OLED	Vertically Stacked OLEDs	-	10.5	14.4	-	-	[83]
2015	OLED	Vertically Stacked OLEDs	-	3.4	36.8	-	(0.44, 0.45)	[20]
2016	OLED	Vertically Stacked OLEDs	1300	-	-	2.5	(0.326, 0.381)	[68]
2017	OLED	Vertically Stacked OLEDs	43,594	12.32	18.1	28.8	(0.38, 0.44)	[62]
2018	OLED	Vertically Stacked OLEDs	-	-	18.1	-	(0.375, 0.395)	[61]
2018	OLED	Vertically Stacked OLEDs	-	49.4	33.4	-	(0.467, 0.423)	[18]
2022	OLED	Vertically Stacked OLEDs	5748.4	15.1	46.3	42.9	(0.247, 0.579)	[67]
2004	OLED	Single Emission Layer	-	12	42	-	(0.43, 045)	[70]
2009	OLED	Single Emission Layer	-	19.3	42.5	52.8	(0.33, 0.39)	[71]
2015	OLED	Single Emission Layer	-	13.6	14.5	22.5	(0.36, 0.38)	[72]
2020	OLED	Single Emission Layer	30,000	23.6	55.1	52.7	(0.34, 0.36)	[73]
2004	OLED	Double Emission Layers	-	19.3	64	-	-	[75]
2010	OLED	Double Emission Layers	-	34	90	-	(0.45, 0.47)	[76]
2016	OLED	Double Emission Layers	-	23.3	63.2	-	(0.433, 0.458)	[59]
2016	OLED	Multiple Emission Layers	11,000	-	40.8	39.8	(0.32, 0.39)	[60]
2018	OLED	Multiple Emission Layers	-	23.1	59.0	-	(0.324, 0.337)	[77]
2019	OLED	Multiple Emission Layers	23,730	-	33.42	32.74	(0.391, 0.471)	[63]
2020	QLED	Side-by-Side Structure	247,000	22.9	-	9.8	(0.16, 0.77)	[34]
2017	QLED	Vertically Stacked QLEDs	4206	2.0	0.46	4.75	(0.30, 0.44)	[80]
2015	QLED	Stacked Emitter Layers	1170	0.6	0.6	0.9	(0.33, 0.253)	[24]
2022	QLED	Stacked Emitter Layers	1000	13.3	-	-	(0.283, 0.305)	[35]
2020	Hybrid LED	Vertically Stacked Structure	107,000	26.02	20.31	-	(0.34, 0.36)	[37]
2022	Hybrid LED	Vertically Stacked Structure	7735	8.6	-	-	(0.67, 0.30)	[36]
2022	Hybrid LED	Vertically Stacked Structure	24,911	13.7	-	-	(0.16, 0.76)	[36]

3. Flexible and Wearable Multi-Color Devices

In the era of artificial intelligence, flexible and wearable devices become more and more popular in our daily life. With the development of comfortable and stretchable electrodes and substrates, flexible displays based on deformable OLEDs and QLEDs have been widely developed. Wearable optoelectronics typically require light sources that can be manufactured as thin films, such as OLEDs and QLEDs. By integrating flexible multi-color LED with stretchable electronic devices, invisible signals such as temperature and heart rate can be converted into visible colors, which could be used in various types of wearable sensors and healthcare systems.

In 2017, Ja Hoon Koo et al. reported a wearable electrocardiogram monitor via integrating sensors, a carbon nanotube signal amplifier, and ultrathin voltage-dependent color-tunable OLEDs (Figure 13a) [6]. Multi-color OLEDs are used for the colorimetric display of the retrieved electrocardiogram signals. By optimizing the structure of multi-OLEDs, the wearable OLEDs exhibit electrocardiogram-dependent color changes from dark red to pale red, to white, to sky blue, and finally to deep blue. The ultrathin design enables the device to conform to the curvilinear and dynamic surface of human skin and exhibit excellent stability and reliability after repeated deformations (Figure 13b,c).

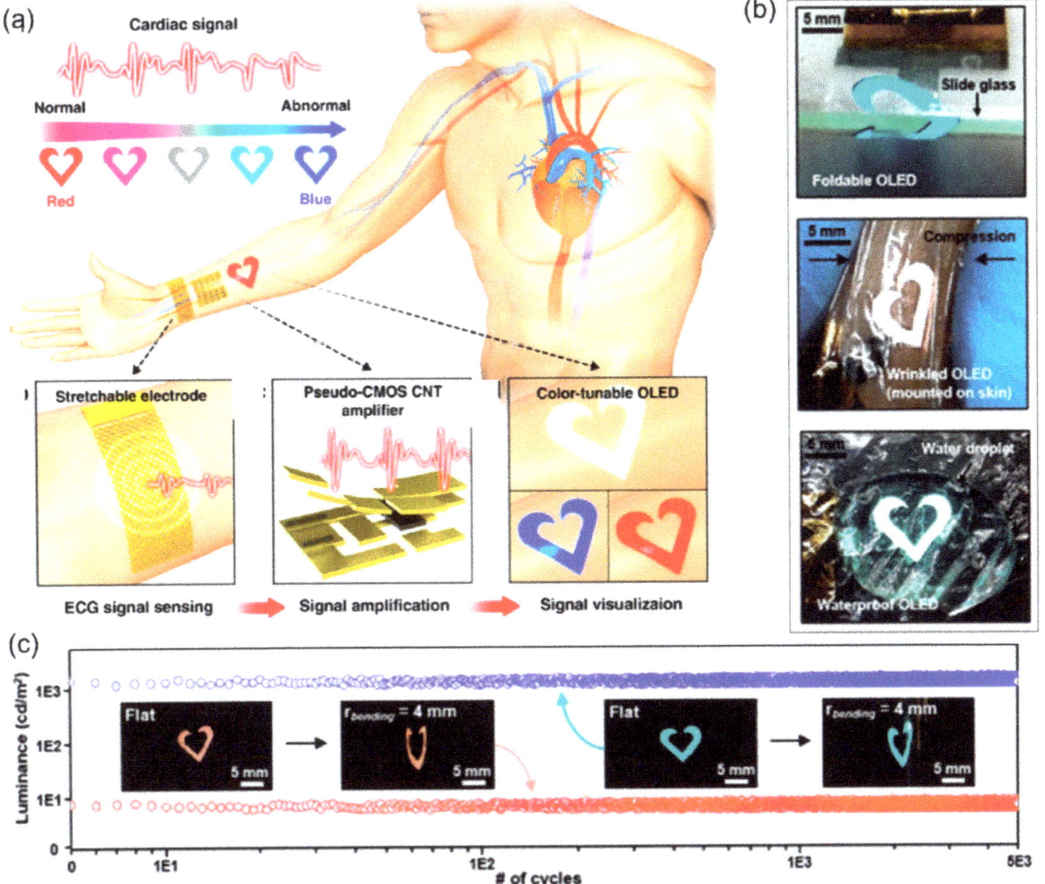

Figure 13. *Cont.*

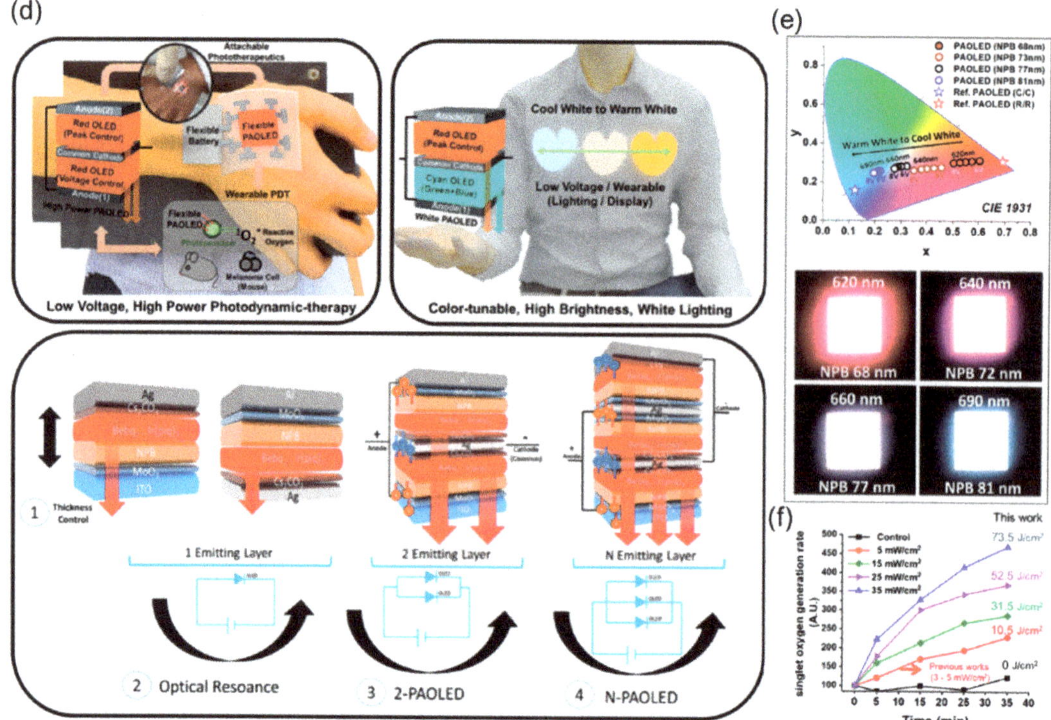

Figure 13. Flexible and Wearable Multi-color Devices. (**a**) Schematic illustration of the wearable cardiac-monitoring system. (**b**) Photographs of the OLEDs after folding along a sliding glass, wrinkled after worn on human skin, and under a water droplet. (**c**) Stable luminance of the blue and red emission during multiple bending experiments (with $r_{bending}$ = 4 mm, 1000 times). Reprinted with permission from Ref. [6]. 2017, ACS Nano. (**d**) Schematic illustration of the red OLED-based wearable PDT system and the color-tunable OLED-based wearable display system. Realization of high power by changing the organic thickness of the ITO reference OLED, using the optical resonance effect of a microcavity reference OLED, through the OLED structural design that electrically stacks an ITO reference OLED and a microcavity reference OLED, and through an OLED structural design that electrically stacks N OLEDs. (**e**) *CIE* 1931 color coordinates and photos of the color-tunable OLED. (**f**) Graph of singlet oxygen generation rate according to the irradiation energy of the OLED. Reprinted with permission from Ref. [7]. 2020, ACS Nano.

Although OLED-based optoelectronic devices have successfully realized wearable applications, current OLED technologies require high voltage and lack the power needed for wearable photodynamic therapy (PDT) applications. In 2020, Yongmin Jeon et al. presented a parallel-stacked multi-color OLED with high power, more than 100 mW cm^{-2}, at low voltage (<8 V) (Figure 13d) [7]. The parallel-stacked OLED with color tuning is realized through OLED color combination, and a high brightness of over 30,000 cd m^{-2} is obtained, below 8.5 V (Figure 13e). Confirming its potential application to PDT, the measured singlet oxygen generation ratio of the parallel-stacked OLED is found to be 3.8 times higher than the reference OLED (Figure 13f).

4. Challenges and Perspectives

Overall, the types, characteristics, and possible wearable applications of multi-color LEDs are introduced in detail. The realization of multi-color LEDs is early achieved by the side-by-side patterning of red, green, and blue sub-pixels onto the pixelated display panel. The devices exhibit good optical performances since the light directly emits from respective units through TFT addressing each pixel independently. However, the fill factors and pixel density are inevitably sacrificed due to the lateral integration configuration. Vertical stacking LED configuration could circumvent the limitation of red/green/blue side-by-side color pixels. Because the full color is attained in a single color-tunable pixel instead of three red/green/blue pixels, the pixel density and fill factor of a display with color-tunable pixels can be enhanced three times. However, because of the light loss and the resistance increase in the semitransparent central electrode, the brightness of multi-color LEDs inevitably decreases, and a high voltage is needed to drive multiple LED units. Stacking various emitter layers with different emission wavelengths in a single LED is a direct and simple way to achieve multi-color emission compared with side-by-side and vertically stacked LED units. However, the transition of the complex exciton region between the two emission layers tends to reduce the device's efficiency.

Therefore, different types of multi-colored LEDs have their respective advantages and disadvantages. Developing particular strengths and circumventing their weaknesses to be applied to appropriate fields is challenging. Although multi-color LED devices have made many remarkable signs of progress in the past few years, there are still some important problems that need further consideration to meet application requirements.

Firstly, the efficiency (such as *EQE*, *PE*, *CE*) of the multi-color LEDs is still too low to meet the needs of practical applications. For example, the *EQE* that gives the ratio of extracted photons over injected charges of most multi-colored LEDs is below 20%. This may be because the efficiency of monochrome LEDs such as blue LEDs still needs to be improved. In addition, the multi-color LED with a complex structure formed by stacking multiple LED units or various emitter layers leads to inadequate recombination of charge carriers. Thus, luminous materials with excellent properties need to be developed and the device structure to achieve an efficient and balanced carrier injection should be optimized. In addition, designing the optical structure to utilize the microcavity effect could further maximize the efficiency of multi-color LEDs.

Secondly, the high-stability device with a long lifetime is critical for practical applications, but few multi-color LEDs focus on it. Therefore, more attention should be paid to the lifetime of multi-color LEDs in future research. A blue LED emitter is the most unstable emitter, and blue LEDs mostly are required to create a full-color LED display. Thus, it is necessary to develop a commercially efficient phosphorescent blue emitter. In addition, ingeniously designing the multi-color device structure employing mature high-stability red and green LEDs so that it does not require a blue emitter to achieve a full-color display is also a solution.

Thirdly, although multi-color LEDs can realize color tunability controlled by the bias voltage, the range of color variation is still limited, and no arbitrary color within full color can be achieved. In addition, the sensitivity of color variation to voltage should also be considered. When sensitive multi-color LEDs and sensors are integrated to construct optoelectronic devices, little changes in signal intensity can achieve significant variations in colors. Therefore, in the future, it is necessary to expand the degree of the color change in multi-color LEDs tuned by bias and improve the sensitivity of LED color changing with voltage.

Finally, the challenge for flexible multi-color LED devices is to ensure that they operate reliably after cycle bending. In addition, encapsulation technology that withstands long-term bending is also critical to practical application. Therefore, more attention should be paid to the stability of flexible multi-color LEDs after cycle bending and developing relevant encapsulation technology.

Author Contributions: Conceptualization, X.T. and G.M.; investigation, S.M., Y.Q. and G.M.; writing—original draft preparation, S.M., Y.Q. and G.M.; writing—review and editing, X.T. and G.M.; supervision, X.T.; project administration, X.T.; funding acquisition, X.T., G.M. and M.C. All authors have read and agreed to the published version of the manuscript.

Funding: This work was funded by the National Key R&D Program of China (2021YFA0717600), National Natural Science Foundation of China (NSFC No. 62035004 and NSFC No. 62105022), and

China Postdoctoral Science Foundation (2022M710396). X.T. is sponsored by the Young Elite Scientists Sponsorship Program by CAST (No. YESS20200163).

Institutional Review Board Statement: Not applicable.

Informed Consent Statement: Not applicable.

Data Availability Statement: Not applicable.

Conflicts of Interest: The authors declare no conflict of interest.

References

1. Chang, M.H.; Das, D.; Varde, P.V.; Pecht, M. Light Emitting Diodes Reliability Review. *Microelectron. Reliab.* **2012**, *52*, 762–782. [CrossRef]
2. Wang, R.; Xiang, H.; Chen, J.; Li, Y.; Zhou, Y.; Choy, W.C.H.; Fan, Z.; Zeng, H. Energy Regulation in White-Light-Emitting Diodes. *ACS Energy Lett.* **2022**, *7*, 2173–2188. [CrossRef]
3. Wu, Y.; Ma, J.; Su, P.; Zhang, L.; Xia, B. Full-Color Realization of Micro-Led Displays. *Nanomaterials* **2020**, *10*, 2482. [CrossRef]
4. Huang, Y.; Hsiang, E.L.; Deng, M.Y.; Wu, S.T. Mini-LED, Micro-LED and OLED Displays: Present Status and Future Perspectives. *Light Sci. Appl.* **2020**, *9*, 105. [CrossRef] [PubMed]
5. Ma, L.; Shao, Y.F. A brief review of innovative strategies towards structure design of practical electronic display device. *J. Cent. South Univ.* **2022**, *27*, 1624–1644. [CrossRef]
6. Koo, J.H.; Jeong, S.; Shim, H.J.; Son, D.; Kim, J.; Kim, D.C.; Choi, S.; Hong, J.I.; Kim, D.H. Wearable Electrocardiogram Monitor Using Carbon Nanotube Electronics and Color-Tunable Organic Light-Emitting Diodes. *ACS Nano* **2017**, *11*, 10032–10041. [CrossRef] [PubMed]
7. Jeon, Y.; Noh, I.; Seo, Y.C.; Han, J.H.; Park, Y.; Cho, E.H.; Choi, K.C. Parallel-Stacked Flexible Organic Light-Emitting Diodes for Wearable Photodynamic Therapeutics and Color-Tunable Optoelectronics. *ACS Nano* **2020**, *14*, 15688–15699. [CrossRef] [PubMed]
8. Mao, M.; Lam, T.L.; To, W.P.; Lao, X.; Liu, W.; Xu, S.; Cheng, G.; Che, C.M. Stable, High-Efficiency Voltage-Dependent Color-Tunable Organic Light-Emitting Diodes with a Single Tetradentate Platinum(II) Emitter Having Long Operational Lifetime. *Adv. Mater.* **2021**, *33*, e2004873. [CrossRef] [PubMed]
9. Liang, C.J.; Choy, W.C.H. Tunable Full-Color Emission of Two-Unit Stacked Organic Light Emitting Diodes with Dual-Metal Intermediate Electrode. *J. Organomet. Chem.* **2009**, *694*, 2712–2716. [CrossRef]
10. Nakamura, K.; Ishikawa, T.; Nishioka, D.; Ushikubo, T.; Kobayashi, N. Color-Tunable Multilayer Organic Light Emitting Diode Composed of DNA Complex and Tris(8-Hydroxyquinolinato)Aluminum. *Appl. Phys. Lett.* **2010**, *97*, 95–98. [CrossRef]
11. Tsuzuki, T.; Tokito, S. Highly Efficient, Low-Voltage Phosphorescent Organic Light-Emitting Diodes Using an Iridium Complex as the Host Material. *Adv. Mater.* **2007**, *19*, 276–280. [CrossRef]
12. Jiang, Y.; Lian, J.; Chen, S.; Kwok, H.S. Fabrication of Color Tunable Organic Light-Emitting Diodes by an Alignment Free Mask Patterning Method. *Org. Electron.* **2013**, *14*, 2001–2006. [CrossRef]
13. Geffroy, B.; le Roy, P.; Prat, C. Organic Light-Emitting Diode (OLED) Technology: Materials, Devices and Display Technologies. *Polym. Int.* **2006**, *55*, 572–582. [CrossRef]
14. Gayral, B. LEDs for Lighting: Basic Physics and Prospects for Energy Savings. *Comptes Rendus Phys.* **2017**, *18*, 453–461. [CrossRef]
15. Kanno, H.; Hamada, Y.; Takahashi, H. Development of OLED with High Stability and Luminance Efficiency by Co-Doping Methods for Full Color Displays. *IEEE J. Sel. Top. Quantum Electron.* **2004**, *10*, 30–36. [CrossRef]
16. Schwartz, G.; Reineke, S.; Rosenow, T.C.; Walzer, K.; Leo, K. Triplet Harvesting in Hybrid White Organic Light-Emitting Diodes. *Adv. Funct. Mater.* **2009**, *19*, 1319–1333. [CrossRef]
17. Müller, C.D.; Falcou, A.; Reckefuss, N.; Rojahn, M.; Wiederhirn, V.; Rudati, P.; Frohne, H.; Nuyken, O.; Becker, H.; Meerholz, K. Multi-Colour Organic Light-Emitting Displays by Solution Processing. *Nature* **2003**, *421*, 829–833. [CrossRef]
18. Park, M.J.; Son, Y.H.; Yang, H.I.; Kim, S.K.; Lampande, R.; Kwon, J.H. Optical Design and Optimization of Highly Efficient Sunlight-like Three-Stacked Warm White Organic Light Emitting Diodes. *ACS Photonics* **2018**, *5*, 655–662. [CrossRef]
19. Guo, F.; Karl, A.; Xue, Q.F.; Tam, K.C.; Forberich, K.; Brabec, C.J. The Fabrication of Color-Tunable Organic Light-Emitting Diode Displays via Solution Processing. *Light Sci. Appl.* **2017**, *6*, e17094. [CrossRef]
20. Fröbel, M.; Schwab, T.; Kliem, M.; Hofmann, S.; Leo, K.; Gather, M.C. Get It White: Color-Tunable AC/DC OLEDs. *Light Sci. Appl.* **2015**, *4*, e247. [CrossRef]
21. Zhang, J.; Ren, B.; Deng, S.; Huang, J.; Jiang, L.; Zhou, D.; Zhang, X.; Zhang, M.; Chen, R.; Yeung, F.; et al. Voltage-Dependent Multicolor Electroluminescent Device Based on Halide Perovskite and Chalcogenide Quantum-Dots Emitters. *Adv. Funct. Mater.* **2020**, *30*, 1907074. [CrossRef]
22. Zhang, Y.; Xie, C.; Su, H.; Liu, J.; Pickering, S.; Wang, Y.; Yu, W.W.; Wang, J.; Wang, Y.; Hahm, J.I.; et al. Employing Heavy Metal-Free Colloidal Quantum Dots in Solution-Processed White Light-Emitting Diodes. *Nano Lett.* **2011**, *11*, 329–332. [CrossRef] [PubMed]
23. Bae, W.K.; Lim, J.; Lee, D.; Park, M.; Lee, H.; Kwak, J.; Char, K.; Lee, C.; Lee, S. R/G/B/Natural White Light Thin Colloidal Quantum Dot-Based Light-Emitting Devices. *Adv. Mater.* **2014**, *26*, 6387–6393. [CrossRef]
24. Kim, J.H.; Lee, K.H.; Kang, H.D.; Park, B.; Hwang, J.Y.; Jang, H.S.; Do, Y.R.; Yang, H. Fabrication of a White Electroluminescent Device Based on Bilayered Yellow and Blue Quantum Dots. *Nanoscale* **2015**, *7*, 5363–5370. [CrossRef] [PubMed]

25. Wepfer, S.; Frohleiks, J.; Hong, A.R.; Jang, H.S.; Bacher, G.; Nannen, E. Solution-Processed $CuInS_2$-Based White QD-LEDs with Mixed Active Layer Architecture. *ACS Appl. Mater. Interfaces* **2017**, *9*, 11224–11230. [CrossRef]
26. Mu, G.; Rao, T.; Zhang, S.; Wen, C.; Chen, M.; Hao, Q.; Tang, X. Ultrasensitive Colloidal Quantum-Dot Upconverters for Extended Short-Wave Infrared. *ACS Appl. Mater. Interfaces* **2022**, *14*, 45553–45561. [CrossRef] [PubMed]
27. Shen, H.; Gao, Q.; Zhang, Y.; Lin, Y.; Lin, Q.; Li, Z.; Chen, L.; Zeng, Z.; Li, X.; Jia, Y.; et al. Visible quantum dot light-emitting diodes with simultaneous high brightness and efficiency. *Nature Photon* **2019**, *13*, 192–197. [CrossRef]
28. Qasim, K.; Zhenbo, Z.; Khatri, N.K.; Xu, Q.; Subramanian, A.; Qing, L.; Wei, L. A Color Tunable Quantum-Dot Light-Emitting Diode Device Driven by Variable Voltage. *J. Nanosci. Nanotechnol.* **2018**, *19*, 1038–1043. [CrossRef]
29. Wang, O.; Wang, L.; Li, Z.; Xu, Q.; Lin, Q.; Wang, H.; Du, Z.; Shen, H.; Li, L.S. High-Efficiency, Deep Blue ZnCdS/Cd: XZn1-XS/ZnS Quantum-Dot-Light-Emitting Devices with an EQE Exceeding 18%. *Nanoscale* **2018**, *10*, 5650–5657. [CrossRef]
30. Li, X.; Lin, Q.; Song, J.; Shen, H.; Zhang, H.; Li, L.S.; Li, X.; Du, Z. Quantum-Dot Light-Emitting Diodes for Outdoor Displays with High Stability at High Brightness. *Adv. Opt. Mater.* **2020**, *8*, 1901145. [CrossRef]
31. Cao, W.; Xiang, C.; Yang, Y.; Chen, Q.; Chen, L.; Yan, X.; Qian, L. Highly Stable QLEDs with Improved Hole Injection via Quantum Dot Structure Tailoring. *Nat. Commun.* **2018**, *9*, 2608. [CrossRef] [PubMed]
32. Kang, H.L.; Kang, J.; Won, J.K.; Jung, S.M.; Kim, J.; Park, C.H.; Ju, B.K.; Kim, M.G.; Park, S.K. Spatial Light Patterning of Full Color Quantum Dot Displays Enabled by Locally Controlled Surface Tailoring. *Adv. Opt. Mater.* **2018**, *6*, 1701335. [CrossRef]
33. Lee, K.H.; Han, C.Y.; Kang, H.D.; Ko, H.; Lee, C.; Lee, J.; Myoung, N.S.; Yim, S.Y.; Yang, H. Highly Efficient, Color-Reproducible Full-Color Electroluminescent Devices Based on Red/Green/Blue Quantum Dot-Mixed Multilayer. *ACS Nano* **2015**, *9*, 10941–10949. [CrossRef] [PubMed]
34. Mei, W.; Zhang, Z.; Zhang, A.; Li, D.; Zhang, X.; Wang, H.; Chen, Z.; Li, Y.; Li, X.; Xu, X. High-Resolution, Full-Color Quantum Dot Light-Emitting Diode Display Fabricated via Photolithography Approach. *Nano Res.* **2020**, *13*, 2485–2491. [CrossRef]
35. Mu, G.; Rao, T.; Chen, M.; Tan, Y.; Hao, Q.; Hao, Q.; Tang, X. Colloidal Quantum-Dot Light Emitting Diodes with Bias-Tunable Color. *Photonics Res.* **2022**, *10*, 1633–1639. [CrossRef]
36. Lee, S.; Hahm, D.; Yoon, S.Y.; Yang, H.; Bae, W.K.; Kwak, J. Quantum-Dot and Organic Hybrid Light-Emitting Diodes Employing a Blue Common Layer for Simple Fabrication of Full-Color Displays. *Nano Res.* **2022**, *15*, 6477–6482. [CrossRef]
37. Zhang, H.; Su, Q.; Chen, S. Quantum-Dot and Organic Hybrid Tandem Light-Emitting Diodes with Multi-Functionality of Full-Color-Tunability and White-Light-Emission. *Nat. Commun.* **2020**, *11*, 2826–2833. [CrossRef]
38. Wood, V.; Bulović, V. Colloidal Quantum Dot Light-Emitting Devices. *Nano Rev.* **2010**, *1*, 5202. [CrossRef]
39. Hong, G.; Gan, X.; Leonhardt, C.; Zhang, Z.; Seibert, J.; Busch, J.M.; Bräse, S. A Brief History of OLEDs—Emitter Development and Industry Milestones. *Adv. Mater.* **2021**, *33*, e2005630. [CrossRef]
40. Pode, R. Organic Light Emitting Diode Devices: An Energy Efficient Solid State Lighting for Applications. *Renew. Sustain. Energy Rev.* **2020**, *133*, 110043. [CrossRef]
41. Marcato, T.; Shih, C.J. Molecular Orientation Effects in Organic Light-Emitting Diodes. *Helv. Chim. Acta* **2019**, *102*, e1900048. [CrossRef]
42. Ingram, G.L.; Lu, Z.-H. Design Principles for Highly Efficient Organic Light-Emitting Diodes. *J. Photonics Energy* **2014**, *4*, 040993. [CrossRef]
43. Shao, J.; Chen, C.; Zhao, W.; Zhang, E.; Ma, W.; Sun, Y.; Chen, P.; Sheng, R. Recent Advances of Interface Exciplex in Organic Light-Emitting Diodes. *Micromachines* **2022**, *13*, 298. [CrossRef] [PubMed]
44. Qasim, K.; Lei, W.; Li, Q. Quantum Dots for Light Emitting Diodes. *J. Nanosci. Nanotechnol.* **2013**, *13*, 3173–3185. [CrossRef] [PubMed]
45. Seok, H.J.; Lee, J.H.; Park, J.H.; Lim, S.H.; Kim, H.K. Transparent Conducting Electrodes for Quantum Dots Light Emitting Diodes. *Isr. J. Chem.* **2019**, *59*, 729–746. [CrossRef]
46. Fukagawa, H. Molecular Design and Device Design to Improve Stabilities of Organic Light-Emitting Diodes. *J. Photopolym. Sci. Technol.* **2018**, *31*, 315–321. [CrossRef]
47. Vidyasagar, C.C.; Muñoz Flores, B.M.; Jiménez-Pérez, V.M.; Gurubasavaraj, P.M. Recent Advances in Boron-Based Schiff Base Derivatives for Organic Light-Emitting Diodes. *Mater. Today Chem.* **2019**, *11*, 133–155. [CrossRef]
48. Sasabe, H.; Kido, J. Recent Progress in Phosphorescent Organic Light-Emitting Devices. *Eur. J. Org. Chem.* **2013**, *2013*, 7653–7663. [CrossRef]
49. Jeong, H.; Shin, H.; Lee, J.; Kim, B.; Park, Y.-I.; Yook, K.S.; An, B.-K.; Park, J. Recent Progress in the Use of Fluorescent and Phosphorescent Organic Compounds for Organic Light-Emitting Diode Lighting. *J. Photonics Energy* **2015**, *5*, 057608. [CrossRef]
50. Shang, Y.; Ning, Z. Colloidal Quantum Dots Surface and Device Structure Engineering for High Performance Light Emitting. *Natl. Sci. Rev.* **2017**, *4*, 107–183. [CrossRef]
51. Xie, B.; Hu, R.; Luo, X. Quantum Dots-Converted Light-Emitting Diodes Packaging for Lighting and Display: Status and Perspectives. *J. Electron. Packag. Trans. ASME* **2016**, *138*, 020803. [CrossRef]
52. Choi, M.K.; Yang, J.; Hyeon, T.; Kim, D.H. Flexible Quantum Dot Light-Emitting Diodes for next-Generation Displays. *npj Flex. Electron.* **2018**, *2*, 1–14. [CrossRef]
53. Shirasaki, Y.; Supran, G.J.; Bawendi, M.G.; Bulović, V. Emergence of Colloidal Quantum-Dot Light-Emitting Technologies. *Nat. Photonics* **2013**, *7*, 13–23. [CrossRef]
54. Chen, F.; Guan, Z.; Tang, A. Nanostructure and Device Architecture Engineering for High-Performance Quantum-Dot Light-Emitting Diodes. *J. Mater. Chem. C* **2018**, *6*, 10958–10981. [CrossRef]
55. Sun, Y.; Jiang, Y.; Sun, X.W.; Zhang, S.; Chen, S. Beyond OLED: Efficient Quantum Dot Light-Emitting Diodes for Display and Lighting Application. *Chem. Rec.* **2019**, *19*, 1729–1752. [CrossRef]

56. Yuan, Q.; Wang, T.; Yu, P.; Zhang, H.; Zhang, H.; Ji, W. A Review on the Electroluminescence Properties of Quantum-Dot Light-Emitting Diodes. *Org. Electron.* **2021**, *90*, 106086. [CrossRef]
57. Niikura, H.; Légaré, F.; Hasbani, R.; Ivanov, M.Y.; Villeneuve, D.M.; Corkum, P.B. Probing Molecular Dynamics with Attosecond Resolution Using Correlated Wave Packet Pairs. *Nature* **2003**, *421*, 826–829. [CrossRef]
58. Wu, S.; Li, S.; Sun, Q.; Huang, C.; Fung, M.K. Highly Efficient White Organic Light-Emitting Diodes with Ultrathin Emissive Layers and a Spacer-Free Structure. *Sci. Rep.* **2016**, *6*, 25821. [CrossRef]
59. Zhuang, X.; Zhang, H.; Ye, K.; Liu, Y.; Wang, Y. Two Host-Dopant Emitting Systems Realizing Four-Color Emission: A Simple and Effective Strategy for Highly Efficient Warm-White Organic Light-Emitting Diodes with High Color-Rendering Index at High Luminance. *ACS Appl. Mater. Interfaces* **2016**, *8*, 11221–11225. [CrossRef]
60. Chen, P.; Chen, B.; Zuo, L.; Duan, Y.; Han, G.; Sheng, R.; Xue, K.; Zhao, Y. High-Efficiency and Superior Color-Stability White Phosphorescent Organic Light-Emitting Diodes Based on Double Mixed-Host Emission Layers. *Org. Electron.* **2016**, *31*, 136–141. [CrossRef]
61. Lee, H.; Cho, H.; Byun, C.-W.; Han, J.-H.; Kwon, B.-H.; Choi, S.; Lee, J.; Cho, N.S. Color-Tunable Organic Light-Emitting Diodes with Vertically Stacked Blue, Green, and Red Colors for Lighting and Display Applications. *Opt. Express* **2018**, *26*, 18351. [CrossRef] [PubMed]
62. Wu, J.Y.; Chen, S.A. Development of a Highly Efficient Hybrid White Organic-Light-Emitting Diode with a Single Emission Layer by Solution Processing. *ACS Appl. Mater. Interfaces* **2018**, *10*, 4851–4859. [CrossRef] [PubMed]
63. Wang, B.; Kou, Z.; Tang, Y.; Yang, F.; Fu, X.; Yuan, Q. High CRI and Stable Spectra White Organic Light-Emitting Diodes with Double Doped Blue Emission Layers and Multiple Ultrathin Phosphorescent Emission Layers by Adjusting the Thickness of Spacer Layer. *Org. Electron.* **2019**, *70*, 149–154. [CrossRef]
64. Kajiyama, Y.; Kajiyama, K.; Aziz, H. Maskless RGB Color Patterning of Vacuum-Deposited Small Molecule OLED Displays by Diffusion of Luminescent Dopant Molecules. *Opt. Express* **2015**, *23*, 16650. [CrossRef]
65. Burrows, P.E.; Forrest, S.R.; Sibley, S.P.; Thompson, M.E. Color-Tunable Organic Light-Emitting Devices. *Appl. Phys. Lett.* **1996**, *69*, 2959–2961. [CrossRef]
66. Zhao, Y.; Chen, R.; Gao, Y.; Leck, K.S.; Yang, X.; Liu, S.; Abiyasa, A.P.; Divayana, Y.; Mutlugun, E.; Tan, S.T.; et al. AC-Driven, Color- and Brightness-Tunable Organic Light-Emitting Diodes Constructed from an Electron Only Device. *Org. Electron.* **2013**, *14*, 3195–3200. [CrossRef]
67. Chang, Q.; Lü, Z.; Yin, Y.; Xiao, J.; Wang, J. Highly Efficient Tandem OLED Based on a Novel Charge Generation Layer of HAT-CN/CuPc Heterojunction. *Displays* **2022**, *75*, 102306. [CrossRef]
68. Cho, S.H.; Kim, E.H.; Jeong, B.; Lee, J.H.; Song, G.; Hwang, I.; Cho, H.; Kim, K.L.; Yu, S.; Kim, R.H.; et al. Solution-Processed Electron-Only Tandem Polymer Light-Emitting Diodes for Broad Wavelength Light Emission. *J. Mater. Chem. C* **2017**, *5*, 110–117. [CrossRef]
69. Kido, J.; Hongawa, K.; Okuyama, K.; Nagai, K. White Light-Emitting Organic Electroluminescent Devices Using the Poly(N-Vinylcarbazole) Emitter Layer Doped with Three Fluorescent Dyes. *Appl. Phys. Lett.* **1994**, *64*, 815–817. [CrossRef]
70. D'Andrade, B.W.; Holmes, R.J.; Forrest, S.R. Efficient Organic Electrophosphorescent White-Light-Emitting Device with a Triple Doped Emissive Layer. *Adv. Mater.* **2004**, *16*, 624–628. [CrossRef]
71. Wang, Q.; Ding, J.; Dongge, M.; Cheng, Y.; Wang, L.; Jing, X.; Wang, F. Harvesting Excitons via Two Parallel Channels for Efficient White Organic LEDs with Nearly 100% Internal Quantum Efficiency: Fabrication and Emission-Mechanism Analysis. *Adv. Funct. Mater.* **2009**, *19*, 84–95. [CrossRef]
72. Huang, M.H.; Lin, W.C.; Fan, C.C.; Wang, Y.S.; Lin, H.W.; Liao, J.L.; Lin, C.H.; Chi, Y. Tunable Chromaticity Stability in Solution-Processed Organic Light Emitting Devices. *Org. Electron.* **2015**, *20*, 36–42. [CrossRef]
73. Ding, D.; Wang, Z.; Li, C.; Zhang, J.; Duan, C.; Wei, Y.; Xu, H. Highly Efficient and Color-Stable Thermally Activated Delayed Fluorescence White Light-Emitting Diodes Featured with Single-Doped Single Emissive Layers. *Adv. Mater.* **2020**, *32*, e1906950. [CrossRef] [PubMed]
74. Zhou, X.; Qin, D.S.; Pfeiffer, M.; Blochwitz-Nimoth, J.; Werner, A.; Drechsel, J.; Maennig, B.; Leo, K.; Bold, M.; Erk, P.; et al. High-Efficiency Electrophosphorescent Organic Light-Emitting Diodes with Double Light-Emitting Layers. *Appl. Phys. Lett.* **2002**, *81*, 4070–4072. [CrossRef]
75. He, G.; Pfeiffer, M.; Leo, K.; Hofmann, M.; Birnstock, J.; Pudzich, R.; Salbeck, J. High-Efficiency and Low-Voltage p-i-n Electrophosphorescent Organic Light-Emitting Diodes with Double-Emission Layers. *Appl. Phys. Lett.* **2004**, *85*, 3911–3913. [CrossRef]
76. Reineke, S.; Lindner, F.; Schwartz, G.; Seidler, N.; Walzer, K.; Lüssem, B.; Leo, K. White Organic Light-Emitting Diodes with Fluorescent Tube Efficiency. *Nature.* **2009**, *459*, 234–238. [CrossRef]
77. Kim, G.W.; Bae, H.W.; Lampande, R.; Ko, I.J.; Park, J.H.; Lee, C.Y.; Kwon, J.H. Highly Efficient Single-Stack Hybrid Cool White OLED Utilizing Blue Thermally Activated Delayed Fluorescent and Yellow Phosphorescent Emitters. *Sci. Rep.* **2018**, *8*, 16263. [CrossRef]
78. Colvin, V.L.; Schlamp, M.C.; Alivisatos, A.P. Light-Emitting Diodes Made from Cadmium Selenide Nanocrystals and a Semiconducting Polymer. *Nature* **1994**, *370*, 354–357. [CrossRef]
79. Kim, T.H.; Cho, K.S.; Lee, E.K.; Lee, S.J.; Chae, J.; Kim, J.W.; Kim, D.H.; Kwon, J.Y.; Amaratunga, G.; Lee, S.Y.; et al. Full-Colour Quantum Dot Displays Fabricated by Transfer Printing. *Nat. Photonics* **2011**, *5*, 176–182. [CrossRef]
80. Zhang, H.; Wang, S.; Sun, X.; Chen, S. All Solution-Processed White Quantum-Dot Light-Emitting Diodes with Three-Unit Tandem Structure. *J. Soc. Inf. Disp.* **2017**, *25*, 143–150. [CrossRef]

81. Ki Bae, W.; Kwak, J.; Lim, J.; Lee, D.; Ki Nam, M.; Char, K.; Lee, C.; Lee, S. Multicolored Light-Emitting Diodes Based on All-Quantum-Dot Multilayer Films Using Layer-by-Layer Assembly Method. *Nano Lett.* **2010**, *10*, 2368–2373. [CrossRef] [PubMed]
82. Wang, T.; Chen, Z.; Zhang, H.; Ji, W. Color-Tunable Alternating-Current Quantum Dot Light-Emitting Devices. *ACS Appl. Mater. Interfaces* **2021**, *13*, 45815–45821. [CrossRef] [PubMed]
83. Joo, C.W.; Moon, J.; Han, J.H.; Huh, J.W.; Lee, J.; Cho, N.S.; Hwang, J.; Chu, H.Y.; Lee, J.I. Color Temperature Tunable White Organic Light-Emitting Diodes. *Org. Electron.* **2014**, *15*, 189–195. [CrossRef]

Review

Infrared-to-Visible Upconversion Devices

Tianyu Rao [1], Menglu Chen [1,2,3], Ge Mu [1,*] and Xin Tang [1,2,3,*]

1 School of Optics and Photonics, Beijing Institute of Technology, Beijing 100081, China; 3120210632@bit.edu.cn (T.R.); menglu@bit.edu.cn (M.C.)
2 Beijing Key Laboratory for Precision Optoelectronic Measurement Instrument and Technology, Beijing Institute of Technology, Beijing 100081, China
3 Yangtze Delta Region Academy, Beijing Institute of Technology, Jiaxing 314019, China
* Correspondence: 7520210145@bit.edu.cn (G.M.); xintang@bit.edu.cn (X.T.)

Abstract: Infrared imaging plays remarkable roles in various fields including military, biomedicine, aerospace, and artificial intelligence. However, traditional infrared imaging systems have plenty of disadvantages such as large volume, high cost, and complex fabrication process. Emerging infrared upconversion imaging devices can directly convert low-energy infrared photons into high-energy visible light photons, thus they are promising to accomplish pixel-less high-resolution infrared imaging at low cost. In this paper, recent advances and progress of infrared-to-visible upconversion devices are summarized. We further offer the main limitations of upconversion technology and the challenges that need to be addressed for the future development of infrared upconverters.

Keywords: infrared-to-visible upconversion devices; infrared imaging; upconversion efficiency

Citation: Rao, T.; Chen, M.; Mu, G.; Tang, X. Infrared-to-Visible Upconversion Devices. *Coatings* **2022**, *12*, 456. https://doi.org/10.3390/coatings12040456

Academic Editor: Alessandro Latini

Received: 28 February 2022
Accepted: 24 March 2022
Published: 27 March 2022

1. Introduction

Human vision is limited in the visible wavelength range of 400 to 700 nm. The detection and visualization of electromagnetic waves with wavelengths beyond the range of human vision, such as visualization of near-infrared (NIR), short-wave infrared (SWIR), or mid-infrared (MIR) light, is of vital importance and challenging. Optical sensing and imaging devices in the NIR have a wide range of applications in communications [1], bio-imaging [2], machine vision systems, non-invasive subsurface vision, night vision [3,4], optical sensors for semiconductor inspection, and non-contact industrial and consumer electronic displays. SWIR with wavelengths ranging from 1.5 to 3 μm covers the second and third biological windows. SWIR photodetection and visualization in this wavelength range plays an essential role in numerous applications [5–7] such as environmental pollution, environmental monitoring, bio-imaging, medicine, agriculture, automobiles, food, health monitoring, etc. These applications are due to SWIR's unique properties, for example, reducing light scattering and absorption key role, deeply penetrating into biological tissues, and reducing phototoxicity. MIR refers to 3–5 μm-band infrared, which belongs to the "atmosphere window", that is, there are many infrared radiation transmission components from the atmosphere. Therefore, an MIR detector [8–10] plays an important role in atmospheric monitoring, gas detection, infrared countermeasures, and other aspects. At the same time, MIR has great research value in the field of military infrared detection, especially infrared thermal imaging [11–13] and infrared guidance.

Infrared imaging systems contain objective information that cannot be reflected by visible light imaging and have indispensable applications in many fields such as military, aerospace, remote sensing mapping, autonomous driving, astronomical science, medical, health care, and others. Focusing on the new generation of cutting-edge application scenarios and demands such as unmanned aerial vehicles, mobile reconnaissance, and smart cities in the future, infrared imaging has become the focus of current research.

However, up to now, the most mature infrared imaging technology is the traditional infrared focal plane imaging technology. This imaging technology mainly has the following two problems: (1) the interconnection of the detector and the readout circuit needs to grow tens of thousands of indium pillars and complete it through the flip-chip interconnection process, which is not only complicated and expensive, but also unreliable; (2) the output electric signal not only needs to be read out by the readout circuit but the imaging processing circuit is also very complicated. It can be seen that the preparation technology of infrared materials and the conversion processing of photoelectric signals have become the core problems that limit the development of infrared imaging technology.

In response to the above challenges, a new type of photonic upconverter called infrared-to-visible upconversion device has emerged. The upconverter can convert invisible infrared images to visible light without pixelated processing in the device. In contrast to traditional infrared imaging technology, upconversion devices convert low-energy incident infrared photons into high-energy output visible light photons, which are then imaged by commercial visible light imaging technology or directly observed by the naked eyes. Currently, there are primarily two upconversion mechanisms: nonlinear optical processes and linear optical processes.

Nonlinear upconversion is the synthesis of infrared light and visible light into a new beam of visible light whose frequency is the sum of the two (sum frequency process). This kind of upconversion device is directly realized by upconversion luminescent materials. Triplet–triplet annihilation (TTA)-based near-infrared to visible molecular photon upconversion [14–22] and nanoparticle-based upconversion [23–26] are two main upconversion luminescent materials. Early achievements on infrared upconversion are mainly based on Ge/GaAs heterojunction structures and rare-earth ion doping systems. There are two representative works. Yoh Mita demonstrated an infrared-excitable phosphor-based upconverter that can detect 1.5 μm NIR and has visible or NIR emissions [27]. Yuhu Wang et al., in 1993, proposed a device whose upconversion situation has been observed with Er^{3+}-doped $BaCl_2$ phosphors [28]. From these nonlinear upconversions mentioned above, the conclusion that can be drawn is that those materials need two or more photons, and then through nonradiative relaxation the photons reach the luminescence level, transition to the ground state, and emit a visible photon. With the purpose of realizing high-efficiency infrared light upconversion, it is necessary for the propagation among infrared light, visible light, and upconverted light to meet the conditions of mutual matching. This nonlinear process based on momentum matching often involves complicated and difficult optical path and material design and usually requires an expensive high-power pulsed light source to emit visible light, which is unfavorable to its wider application.

Therefore, an infrared-to-visible upconversion device based on linear photon upconversion is an ideal scheme to realize high-performance, low-cost infrared imaging. Its working principle is that by integrating infrared photodetectors (PD) that absorb infrared photons and the light-emitting diodes (LED) that excite visible photons together, the low-energy infrared light can be converted into high-energy visible light by using the linear conversion process of light–electricity–light, thus realizing infrared detection and imaging. This infrared upconversion imaging method has many advantages. Firstly, there is no need to pixelize the large-area imaging device and interconnect and align indium pillars for each pixel by a flip-chip bonding process, which greatly simplifies the process. Secondly, visible light images can be directly observed with the naked eye without complicated readout circuits and signal processing processes, and infrared imaging can be realized by using efficient visible light imaging technologies such as CCD cameras, which can significantly improve imaging efficiency, resolution, and sensitivity; Thirdly, in the later stage, there is no need to carry out complicated pseudo-color-coding processing on the image. The pseudo-color images, which are more sensitive for the naked eye, can be directly realized from hardware, for the multi-color LED that can produce the color change is coupled with the infrared detector.

Based on the difference in semiconductor materials of PD and LED units, in this article, the type of optical upconverters can be divided into four categories: all-inorganic upconversion devices, all-organic upconversion devices, inorganic semiconductor infrared detector and organic LED (OLED) hybrid optical upconverter (hybrid organic–inorganic upconversion devices) and colloidal quantum dots-based upconversion devices. According to the different infrared bands detected by the four kinds of upconverters, they are discussed separately. Progress in upconversion devices is summarized in Figure 1.

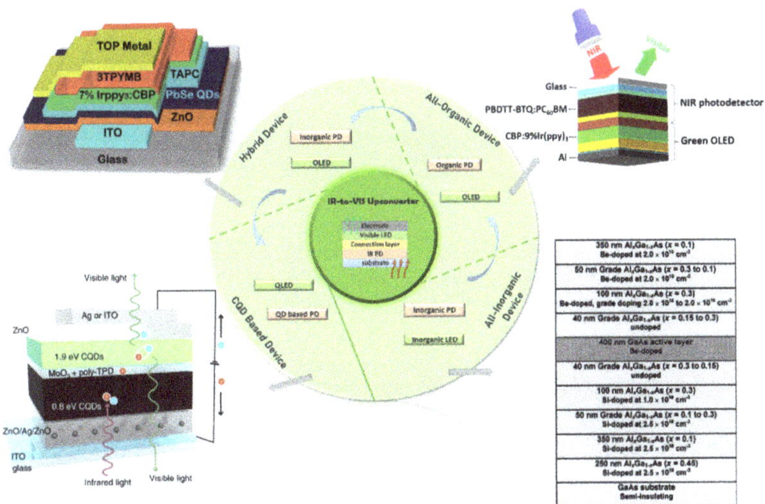

Figure 1. Progress in upconversion devices. The deep green circle in the center shows the representative structure of upconverters, the light green circle around it demonstrates the main four categories of upconversion devices. The boxes in the light circle show the key differences of different kinds of upconversion devices. These blue arrows in the light circle represent the evolution of upconversion devices. The peripheral pictures are some representative structures of these four kinds of upconverters.

2. Types of Upconversion Device

Compared with traditional infrared imaging technology, infrared upconversion devices have obvious advantages in reducing production cost, simplifying the manufacturing process, improving imaging resolution and sensitivity, reducing the size and weight of the imaging system, etc., and have shown excellent performance in the field of infrared imaging. There have been many reports on infrared upconversion devices both domestically and abroad.

2.1. All-Inorganic Upconversion Devices

In 1995, HC Liu successfully used epitaxial growth technology to connect quantum well infrared detectors (GaAs/AlGaAs) and LEDs in series, and integrated inorganic upconversion devices (Figure 2a,b), which realized infrared upconversion imaging from NIR of ~1.5 μm to shorter wavelengths (~0.9 μm) for the first time [29]. In 2000, Liu et al., proposed a mechanism for upconversion using semiconductor structures. InGaA/InP PD and InAsP/InP light-emitting diodes (LED) were grown on InP substrates to prepare room temperature NIR upconverters. However, the upconversion efficiency was less than 0.0005 W/W due to the energy barrier limiting quantum efficiency within the LED and poor carrier constraints in the InAsP/InP LEDs active region [30].

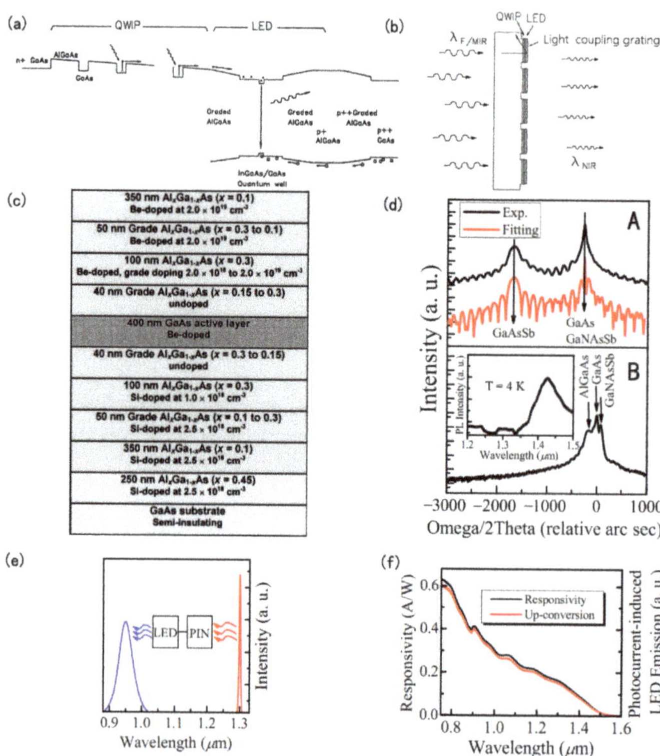

Figure 2. Structures and performance characterization of all-inorganic upconversion devices. (**a**) Bandedge profile of the proposed integrated quantum well intersubband photodetector (QWIP) and LED. (**b**) Schematic geometry of the proposed focal plane array of QWIP-LEDs [29]. Copyright 1955, Electronics Letters. (**c**) Schematic of experimental setup [31]. Copyright 2004, Journal of Applied Physics. (**d**) Picture A shows the measured and fitted X-ray rocking curves for 100 nm thick GaAsSb/GaNAsSb two-layer structure on GaAs substrate. Picture B shows the measured X-ray rocking curve for GaNAsSb layer on GaAs substrate. The inset of picture B shows the PL spectrum of the GaNAsSb layer at 4K in a low-temperature environment. (**e**) Schematic of the test setup structures. The curves also present the intensity of absorbed light and emitted light. (**f**) PD's responsivity under room temperature (black line) and the photocurrent-induced LED light emission of the upconversion device (red line) [32]. Copyright 2009, Applied Physics Letters.

In 2004, to address the mismatch between the GaAs substrate and the active layer lattice of the GaAs detector, which increases rapidly with the increase of indium content, InGaAs/InP PD and GaAs/AlGaAs LED were combined by wafer fusion (Figure 2c). This processing technique was used to integrate heterogeneous semiconductor materials regardless of their lattice mismatch. The results show that this method is effective and improves the efficiency of the upconversion (~0.018 W/W) [31]. However, for pure inorganic upconverter, even if the upconverter is integrated by wafer fusion technology, the cost of epitaxial growth cannot be further reduced, and the ideal conversion efficiency cannot be obtained. Based on a similar structure, in 2009, Shanghai Jiaotong University reported an inorganic upconversion device integrated with GaAsSb/GaAs PD and GaAs/AlGaAs LED (Figure 2d–f), which upconverted NIR light of about 1.3 μm to about 0.95 μm NIR light, with an upconversion efficiency of 4.8% [32].

However, due to the very large energy band difference between LED and infrared photodetection materials, the device can only achieve upconversion from longer wavelengths to shorter wavelengths in the infrared range, not visible light. In addition, the epitaxial growth of traditional upconverters requires very high lattice parameter matching.

2.2. All-Organic Upconversion Devices

Organic electronic devices are expected to be applied to infrared upconversion devices to realize upconversion in the visible light band due to the no lattice matching growth requirements of organic materials. Besides, organic materials are independent of the properties of the substrate material and can be deposited on any material with a low production cost.

All-organic upconverters were first reported on the basis of fluorescent LEDs, which consist of an organic PD sensitive to the infrared and an OLED connected in series. In addition, OLEDs can work at ultra-low temperatures, which ensures that they can work normally when integrated with other detectors in any infrared band. All-organic upconverters promise multiple advantages of organic electronics, and they also provide the possibility of converting NIR scenes into visible images by using large-area devices. These devices can be implemented on flexible substrates at low cost, simplifying the manufacturing process, reducing the production cost, and are currently a commonly used infrared-to-visible light upconversion device.

2.2.1. NIR-to-Visible Upconversion

In 2007, Lu et al. firstly proposed an all-organic upconversion device using 2,4,7-trinitro-9-flfluorenylidene malonitrile (TNFDM) as an infrared photosensitive material, which realized the upconversion process of the visible light emission intensity changing with the incident infrared power and confirmed the feasibility of all-organic NIR–visible light upconversion device [33]. In 2010, Kim et al. prepared an all-organic upconversion device with a conversion efficiency of 2.7% by using SnPc:C_{60} as an infrared detection material and a phosphorescent material with a higher external quantum efficiency than fluorescent materials as a luminescent material [34]. In 2018, D. Yang et al. proposed the concept of using tandem OLED (Figure 3a) that integrates an NIR polymer PD and a green tandem OLED together to fabricate organic upconversion devices, for the use of two layers of light-emitting. The injected electron–hole pairs can be more efficiently converted into visible light than those devices that only have one layer. In this work, they reach one of the highest peak photon-to-photon conversion efficiency up to 29.6% at that time [35]. S. Liu and co-workers, in 2015, demonstrated a transparent upconverter, the NIR light can be converted into visible light with the upconversion device. Moreover, this work is the first one that demonstrated night vision by using a reflective and nonreflective object (Figure 3b) [36].

OLEDs can already achieve the emission of the three primary colors of red, green, and blue, so through the fusion of the three primary colors, they can achieve display in the full visible range. According to the full-color OLED, in 2017, Tachibana et al. reported high-performance all-organic full-color upconversion devices (Figure 3c). They used thermally activated delayed fluorescence (TADF) for the first time. Their result has great potential for NIR imaging applications with large area, low cost, and pixel-less imaging [37].

Upconversion efficiency is vital to upconversion devices. Many attempts have been made in order to improve the photon-to-photon upconversion efficiency. In 2018, Song and his co-workers promoted integrating a photo multiplying organic NIR PD with a high-efficiency TADF-OLED (Figure 3d), and through this method, the photon-to-photon upconversion efficiency can run up to 256% [38].

In these cases cited above, there is no more than 900 nm NIR light upconverted to visible light. In 2018, Strassel and his co-workers extended the NIR spectral peak sensitivity up to 1000 nm, meanwhile, this device is visibly transparent and the innovative use of NIR selective squaraine dye is the highlight of this work [41]. In 2019, the same team reported a solution-processed upconversion device, by adding an electrolyte to the super yellow (SY) emitter (Figure 3e), thereby transforming the OLED into a light-emitting electrochemical cell (LEC), which is more fault-tolerant and has a lower driving voltage than traditional OLED [39]. In the same year, Yeddu et al. raised infrared sensitivity up to 1.1 µm. For such long-wavelength infrared light detection, the detector requires low bandgap organic semiconductors, therefore, they synthesized a novel low bandgap polymer poly 4-(4,8-bis (5-(2-butyloctyl) thiophen-2-yl) benzo [1,2-b:4,5-b′] dithiophen-2-yl)-6,7-diethyl- [1,2,5] thiadiazol [3,4-g] quinoxaline (PBDTT-BTQ) with strong photoresponse in NIR wavelengths of 700−1100 nm which the previously used Si-based PDs cannot offer. PBDTT-BTQ's infrared absorption is superior to others as shown in Figure 3f [40].

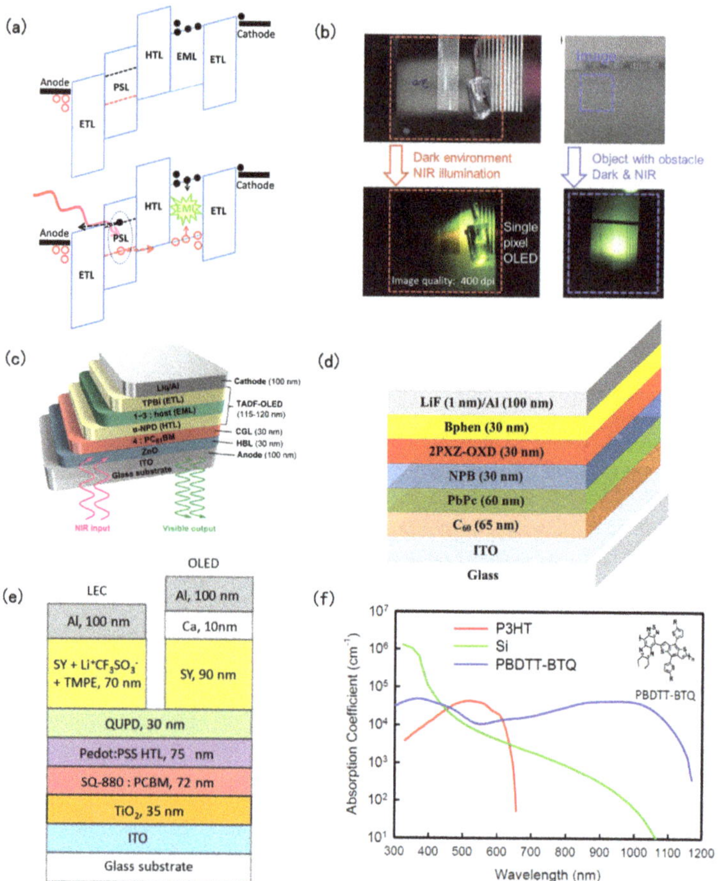

Figure 3. Structures and performance characterization of all-organic NIR-to-visible upconversion devices. (**a**) Structure and working mechanism of the upconversion device under dark (top image) and illumination condition (lower image) [35]. Copyright 2018, Materials Horizons. (**b**) Picture of the line-shadow mask and LED in night vision (left picture), and the picture of the line-shaped shadow mask when an obstacle (waste-water-filled cloudy box) was placed between the mask and the camera (right picture) [36]. Copyright 2015, Advanced Material. (**c**) Structure of the upconversion devices based on TADF-OLEDs [37]. Copyright 2017, ACS Photonics. (**d**) Structure of the upconversion device [38]. Copyright 2018, The Journal of Physical Chemistry Letters. (**e**) Structure of the upconversion device that stacks with a LEC or an OLED as visible light-emitting component [39]. Copyright 2019, ACS Applied Materials & Interfaces. (**f**) Comparison of absorption coefficients of Si, P3HT (film), and PBDTT-BTQ (film). Inset shows the molecular structure of PBDTT-BTQ [40]. Copyright 2019, ACS Photonics.

2.2.2. SWIR-to-Visible Upconversion

In 2020, Li and his co-workers reported their efforts in all solution-based process SWIR-to-visible upconversion devices. They fabricated a blend of a diketopyrrolopyrrole–dithienylthieno[3,2-b] thiophene (DPP–DTT) polymer donor and a hetero-polycyclic aromatic compound SWIR dye with a $CsPbBr_3$ perovskite LED. This device can detect infrared light up to 1400 nm and convert it to 516 nm visible green light [42]. The spectral detection of hybrid upconverters is expanded to the SWIR range for the first time. This advanced device can be operated without cooling and was proved to apply in biological fields, such as providing reliable heart rate monitoring. The structure and performance of this upconverter are shown in Figure 4.

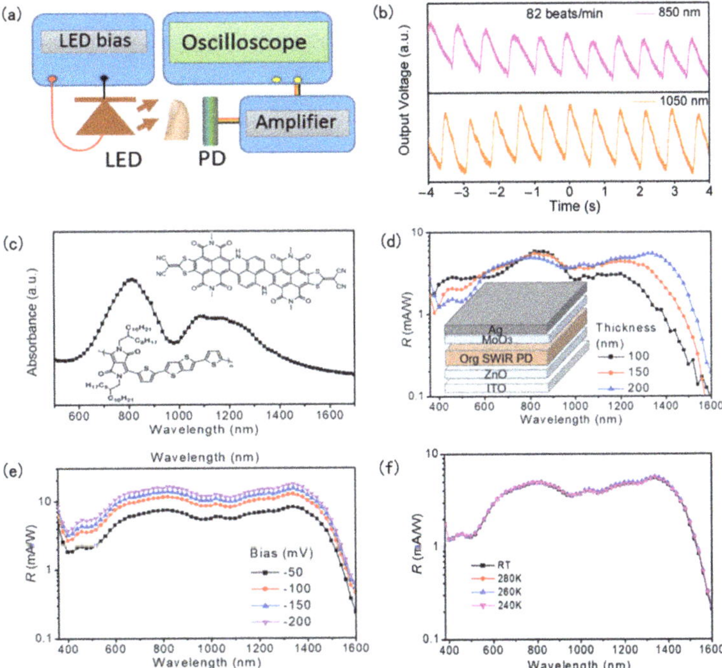

Figure 4. Structure and performance characterization of all-organic SWIR-to-visible upconversion devices. (**a**) A schematic of the measurement setup. (**b**) The heartbeat waveform measured by the device using NIR (850 nm) and SWIR (1050 nm) light illuminations, this picture shows a heart rate of 82 beats min^{-1}. (**c**) The absorption spectrum of the DPP–DTT: SWIR dye blend layer deposited on glass. The insets in (**c**) are molecular structures of the DPP–DTT polymer and the SWIR dye. (**d**) R(λ) of the organic SWIR PDs, with different BHJ layer thicknesses, as a function of the wavelength. The inset in (**d**) is the structure of this organic SWIR PD. (**e**) R(λ) of an organic SWIR PD as a function of the wavelength with a 200 nm thick BHJ layer operated under different reverse biases. (**f**) R(λ) of an organic SWIR PD, with a 200 nm thick BHJ, measured at different temperatures [42]. Copyright 2020, Advanced Science.

In summary, the infrared-to-visible upconversion devices have developed from the traditional all-inorganic structure to the widely used all-organic structure at present, successfully achieving low-cost infrared-to-visible light upconversion imaging. However, for almost every all-organic upconversion device only the NIR band can be detected, and the upconversion efficiency of the device is very low. Generally, it can only work under the high-power infrared laser, which cannot meet the requirements of practical application.

2.3. Hybrid Organic–Inorganic Upconversion Devices

Previously, blended organic and inorganic materials were synthesized and manufactured to form devices. The flexibility of organic chemistry allows organic molecules to form with useful luminescent and conducting properties. Due to the stronger covalent and ionic bonds, inorganic materials can be designed to have better thermal/mechanical stability and higher electromigration. Using different emissive organic materials, the wavelengths of OLED emissions can vary across the full visible spectrum. However, inefficient photo carrier injection from inorganic detectors to OLEDs results in low infrared-to-visible upconversion efficiency. Therefore, there is still a lot of work for researchers to do in order to enhance devices' performance.

2.3.1. NIR-to-Visible Upconversion

In 2008, Chen et al., of the University of Waterloo, added a Ti/Au thin film between the InGaAs infrared detector and the OLED, and the conversion efficiency of the integrated upconversion device was improved to 1.5% [43]. Based on this research, in 2010, Chen et al. obtained an optical upconverter by optimizing the structure of an InGaAs/InP heterojunction phototransistor (HPT) with a gain mechanism and integrating it with an OLED [44]. The structure and equivalent circuit diagram of this upconverter are shown in Figure 5a. The optical upconverter not only realizes the conversion of 1.5 μm NIR to green light (520 nm) at room temperature but also achieves the optical power conversion efficiency of 1.55 W/W (corresponding to an external quantum efficiency of 59%).

In 2012, Chen et al., proposed a pixel-less imaging upconversion device, they employed Technology Computer-Aided Design software (TCAD) to simulate the propagation of photo-generated hole carriers in the upconverter, it was found that no matter how large the size of the incident infrared beam is, the concentration distribution of photo-generated holes at the intermediate interface of i-InGaAs/C_{60} shows that the lateral propagation distance is less than 1 μm [45]. The NIR-to-visible light upconverter was used for pixel-less imaging. The image displayed under the NIR illumination through the shaped hole and the characteristic length of the shaped hole and the image are shown in Figure 5b. From the image they demonstrated, the actual resolution of the device can be calculated. The resulting spatial resolution of 6 μm is smaller than its calculated theoretical value of 1 μm. According to their analysis, the small resolution was mainly due to the limited resolution of digital cameras (~6 μm). However, the infrared pixel-less imaging device achieved a spatial resolution better than 6 μm, which was sufficient for practical applications.

Generally, due to the serial device architecture, the turn-on voltage of the upconversion devices is high. In 2018, Yu et al. reported a low turn-on voltage upconverter, the photocarrier can be efficiently injected into the OLED from the perovskite layer under NIR light, due to the perovskite layer supplied extra voltage to activate the OLED at a 1.9 eV sub-band voltage. On this account, the upconversion device lights up at 1.9 V [46], which is lower than the bandwidth voltage of the emitting layer in the OLED. Further on, Yu et al. demonstrated the pixel-free NIR imaging at a low operation voltage of 3 V which is shown in Figure 5c.

In the same year, Ding et al. reported an NIR-to-visible upconversion strategy based on fully integrated microscale optoelectronic devices, and the dimension of this device is 220 μm × 220 μm (Figure 5d). These devices were implanted in behaving animals, and their effectiveness in photogenetic nerve regulation was proved by in vitro and in vivo experiments. They proposed the theory of the photon–"free electron"–photon process eliminating the limitation of traditional nonlinear upconversion methods [47]. In 2019, Shi et al. reported an ultrafast and low-power optoelectronic NIR-to-visible upconversion device, similar to Ding's theory, they applied the photo–"free electron"–photon concept. Their devices can be excited at low power (1–100 mW/cm^2) and the devices' luminescence decay lifetime (Figure 5e) can be adjusted from ~20 to ~200 ns. Such unique characteristic is beneficial for the application of high-flow chemical detection and dynamic biological stimulations [48].

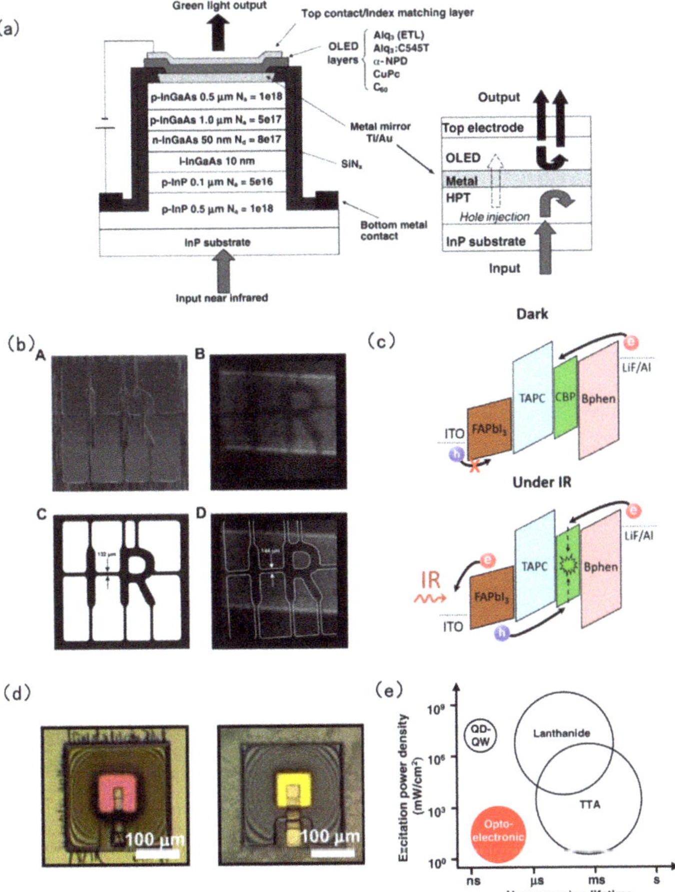

Figure 5. Structures and performance characterization of hybrid organic–inorganic NIR-to-visible upconversion devices. (**a**) Structure of the hybrid upconversion device with integrated InGaAs/InP HPT and OLED. The right picture demonstrates a highly reflective metal layer that inserts in the upconverter, and the embedded optical mirror could improve the absorption and emission efficiency [44]. Copyright 2010, Advanced Materials. (**b**) Picture A of the shaped aperture showing the letters "IR". Picture B of the operating device at 10 V with NIR illumination through the aperture. Picture C is the shape aperture design showing the minimal feature size. Image D of the operating device is overlaid with the shape of the aperture indicating the minimum size of the captured function [45]. Copyright 2012, Advanced Material. (**c**) Band diagram and working mechanism of the upconversion device under the dark and under NIR illumination [46]. Copyright 2018, ACS Applied Materials & Interfaces. (**d**) Microscopic image of upconversion devices with red (**left**) and yellow (**right**) emissions under IR excitation [47]. Copyright 2018, PANs. (**e**) Lifetimes and typical excitation power densities of representative upconversion mechanisms, including lanthanide-based, TTA-based, quantum dot–quantum well-based, and optoelectronic-device-based upconversion designs of this reference [48]. Copyright 2019, Photonics Research.

2.3.2. SWIR-to-Visible Upconversion

In 2007, Ban et al. took the lead in proposing an organic–inorganic hybrid upconversion device integrating in series with an InGaAs/InP infrared detector and an OLED and successfully upconverted SWIR light of 1.55 µm that can be converted into green visible light of 520 nm. However, the conversion efficiency of this infrared-to-visible upconversion device is only 0.7% [49]. The structure and performance of this upconverter are shown in Figure 6a,b.

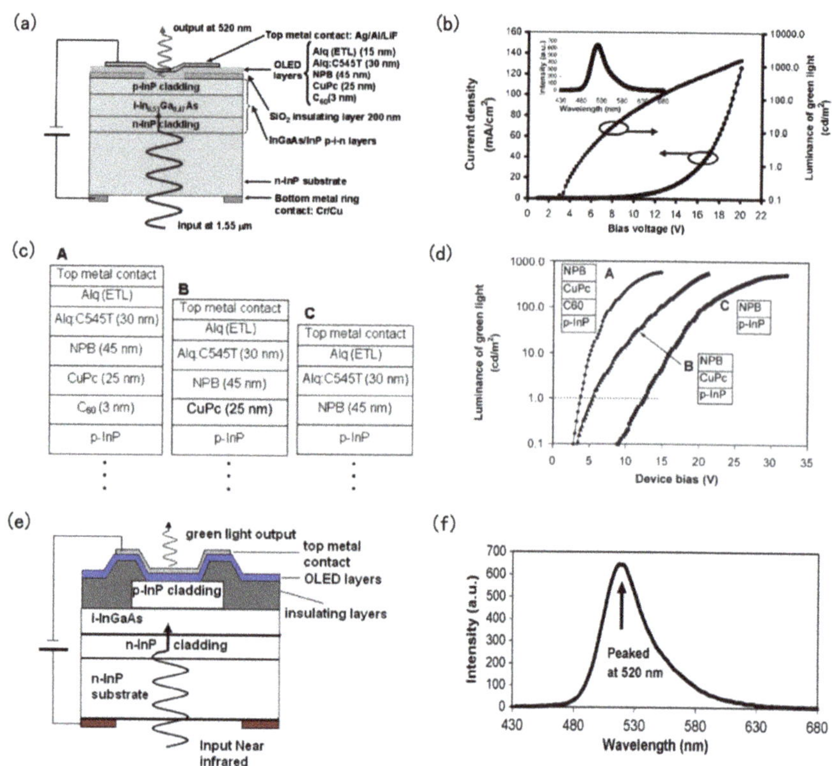

Figure 6. Structures and performance characterization of hybrid organic–inorganic SWIR-to-visible upconversion devices. (**a**) Structure of the inorganic–organic upconversion device. (**b**) Current–voltage–luminance curves of the upconversion device under 1.5 μm infrared illumination (input power density of 1.1 mW/mm^2). The inset shows the emission spectrum from the OLED of the integrated upconversion device with a wavelength peak at 520 nm (green light) [49]. Copyright 2007, Applied Physics Letters. (**c**) Schematic diagram of the three configurations of the hole transform layer (HTL) at the semiconductor–organic interface of the integrated devices. (**d**) OLED green emission luminance of the device A, B, C as a function of the device bias. (**e**) Schematic diagram of the cross-section of one organic–inorganic upconverter device. (**f**) Emission spectrum from the OLED of the integrated upconverter, the peak wavelength of this OLED is 520 nm (green light) [50]. Copyright 2007, IEEE Transactions on Electron Devices.

For an OLED device, its structure is generally composed of an anode, hole injection layer (HIL), hole transport layer (HTL), emission layer, electron transport layer (ETL), electron injection layer (EIL), and cathode. Obviously, in addition to the light-emitting layer, other parts will also affect the performance of the device. Ban et al. also compared the effects of different HTL layer structures on the device performance as shown in Figure 6c. The HTL layer of device C has only one layer of N, N′-di(naphthalene-l-yl)-N, N′-diphenyl-benzidine (NPB), device B adds a layer of CuPc under NPB, and device A continues to add a layer of C_{60} under CuPc [50]. It can be seen from Figure 6d that the organic–inorganic interface has a great influence on the efficient injection of holes, and the turn-on voltage of device C is different from that of devices A and B, because the addition of C_{60} can reduce the carrier injection barrier. The structure and emission spectrum of this upconverter are shown in Figure 6e,f.

2.4. Colloidal Quantum Dots-Based Upconversion Devices

From what was discussed above, most all-organic and hybrid upconversion devices can only convert NIR light into visible light, but usually cannot detect short-wave infrared light, and the application range is severely limited. Besides, the upconversion efficiency of the device is very low [29,37,40–42,45,49], for these organic–inorganic upconversion devices, which are limited by the low photon–electron conversion efficiency of the infrared PD part. Generally, infrared-to-visible upconversion devices can only work under the high-power infrared laser, which cannot meet the practical application requirements. Most of the reported infrared upconversion devices need to be fabricated by high-cost vacuum deposition methods [35,37,45,46], and as for inorganic infrared detectors in hybrid upconversion devices, they still need to be fabricated by epitaxial growth, both of these two problems make the fabrication cost of organic–inorganic hybrid upconversion devices still expensive and unsuitable for large-area imaging applications. In addition, the reported infrared upconversion devices have high operating voltages, which are not conducive to the preparation of flexible devices.

Quantum dots (QDs) have narrow emission linewidth and adjustable bandgap, so that tunable infrared response can be realized in a wide spectral range, and it is the reason that QDs have great potential value in developing quantum information and opto-electronic devices. For example, bulk mercury telluride (HgTe) materials are semi-metallic materials with zero bandgap and large exciton Bohr radii. With the help of the "quantum confinement effect", HgTe QDs [8–10,51–56] can theoretically achieve full spectrum coverage in the infrared to terahertz range. Therefore, the application of colloidal QDs (CQDs), in the field of infrared detectors can realize infrared detection in all major infrared bands of NIR, SWIR, and MIR. In addition, the luminescence spectrum of CQDs is extremely narrow, the color saturation and purity are high, and the optical stability is very good, which can be synthesized in a large-scale liquid phase. Therefore, quantum dot LEDs (QLEDs) have excellent performances of a wide color gamut, long life, and low cost.

CQDs, in particular, can be synthesized by solution treatment, and used in photovoltaics, luminescence, and photodetection. Except for the top electrode, the entire infrared upconversion device is prepared by solution method, which greatly simplifies the preparation of the device. Therefore, the infrared upconversion devices based on CQDs, benefiting from the property of CQDs, can be processed by solution method and can be fabricated into flexible devices.

2.4.1. NIR-to-Visible Upconversion

In 2011, Kim and his co-workers reported their achievement of a low-cost upconversion device with infrared sensitivity up to 1.5 μm using PbSe inorganic colloidal nanocrystals as infrared PD. It was the first time that the NIR-to-visible light converter of CQDs' integrated phosphorescent emitting OLED was reported, the structure and schematic diagram are shown in Figure 7a,b. It must be known that all-organic upconversion devices before this work nearly had no infrared sensitivity beyond 1 μm, moreover, there were not any hybrid devices that could reach a 1.3% photon-to-photon conversion efficiency (that this work obtained) in the past. In contrast to previous studies, Kim et al. employed a wide bandgap ZnO layer as a hole blocking layer to improve the on/off ratio of the device, and the dark current of the device was obviously reduced, moreover the signal-to-noise ratio of the device was increased [57].

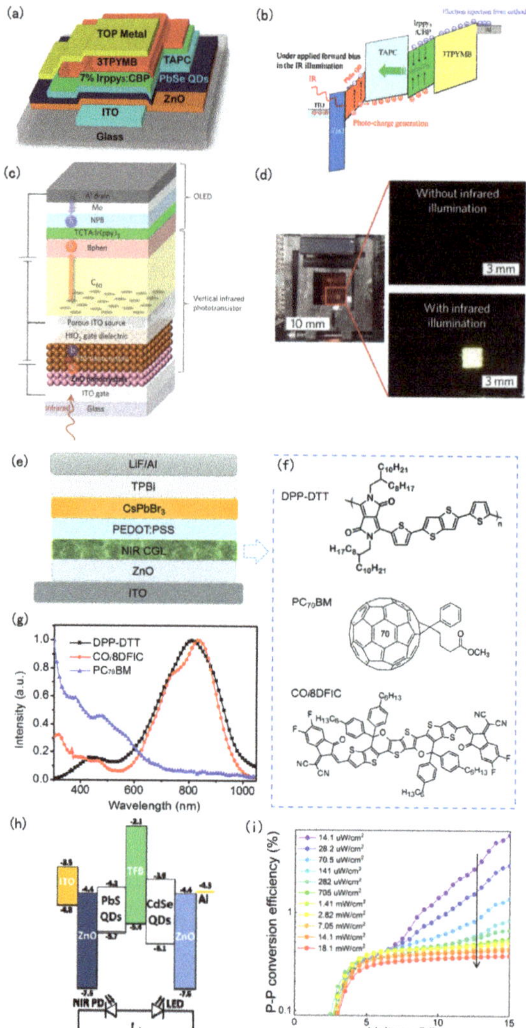

Figure 7. Structures and performance characterization of CQDs based NIR-to-visible upconversion devices. (**a**) Structure of PbSe QD infrared-to-green light upconversion device. (**b**) Schematic energy band diagrams of PbSe QD upconversion devices in the IR illumination [57]. Copyright 2011, Nano Letters. (**c**) Structure of the upconversion device. As shown in the diagram, the colloidal nanocrystals PbS are used as a gate sensitized to the infrared, and an ITO porous layer is used as source electrode of the phototransistor. (**d**) Photo of the sample attached to the measuring case in ambient light (left), and photos of the area of the device (0.04 cm^2) with and without infrared lighting (λ = 940 nm), absent ambient light (right) [58]. Copyright 2016, Nature Photonics. (**e**) Diagram showing the cross-section view of a NIR–visible light upconversion device, with a NIR photodiode and a visible light LED based on $CsPbBr_3$ perovskite. (**f**) Molecular structures of DPP–DTT, $PC_{70}BM$, and CO_i8DFIC, and (**g**) their corresponding normalized absorbance spectra [59]. Copyright 2018, Advanced Optical Materials. (**h**) The structure of the conversion system and circuit show, schematically, the principle of operation of this upconversion device. (**i**) The photon-to-photon conversion efficiency of the device, under different power densities of NIR, as a function of the applied voltage [60]. Copyright 2020, IEEE Access.

In 2016, Yu et al. reported a novel upconversion light-emitting phototransistor (LEPT) by incorporating a phosphorescent OLED into a phototransistor. To be different from the ascending upconverters with two terminals, this LEPT is a vertical three-terminal phototransistor with an infrared photoactive gate integrated into an OLED, which has high-efficiency such as its external quantum efficiency (EQE) of up to $1 \times 10^5\%$ and detectivity of 1.23×10^{13} Jones [58]. The structure of this LEPT is demonstrated in Figure 7c, and a photograph of the sample clamped in the measurement box is shown in Figure 7d. In 2018, Li et al. demonstrated a perovskite LED based on $CsPbBr_3$ with high performance, it can be processed in solution and has a narrow emission spectrum. Their research proved that the NIR-sensitive polymer DPP–DTT blend with a norfullerene acceptor COi8DFIC is an appropriate layer of charge generation layer for the upconversion procession, as a result of its unique combination of high NIR absorption and low visible light absorption (Figure 7e–g) [59]. In 2020, Tang and his co-workers proposed a high conversion efficiency for all QDs-based upconverters (Figure 7h). Former upconversion systems were always manufactured by inorganic semiconductors using wafer fusion technology. However, they were limited by the mismatch of lattice between different semiconductors, leading to poor photonic conversion efficiency. As for the all-organic devices, their efficiency has been improved a lot, but the stability of these upconverters is still a potential issue for their application. In this reference, the use of innately stable, efficient, and high-performance QDs made the device achieve a high photon-to-photon conversion efficiency of 6.3% (Figure 7i), and can be fabricated by solution process [60].

2.4.2. SWIR-to-Visible Upconversion

In 2020, Ning Zhijun's research group from ShanghaiTech University integrated CQD infrared detectors and colloidal QLED for the first time to obtain a CQD infrared-to-visible light upconversion device (Figure 8a). This all-CQDs upconversion device had high photonic conversion efficiency of up to 6.5%, and it can be turned on at a low applied bias of only 2.5 V. Through this upconverter the short-wave infrared (Figure 8b) with a power density of 10–600 $mW \cdot cm^{-2}$ was converted into visible green light (520 nm) (Figure 8c). Ning's group incorporated silver nanoparticles in the PD's layers that can extract charges (Figure 8d) with the aim of optimizing the carrier tunneling and making the PD layer provide enough photocurrent so that the LED can be turned on successfully. This advanced achievement confirms the application of CQDs in upconversion devices and the feasibility of realizing SWIR upconversion imaging [61].

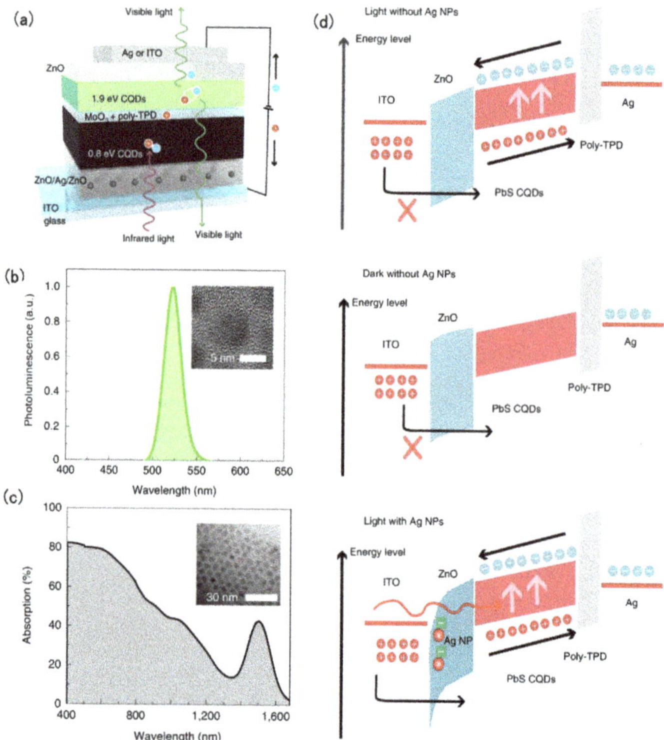

Figure 8. Structures and performance characterization of CQDs based SWIR-to-visible upconversion devices. (**a**) Structure and composition of upconversion devices. (**b**) Photoluminescence spectra of CdSe/ZnS core/shell QDs, with the peak position at 525 nm. Inset: transmission electron microscopy (TEM) of CdSe/ZnS QDs. (**c**) Absorbance of PbS QD thin films, with the exciton peak position at 1500 nm. Inset: TEM of PbS QDs. (**d**) PD energy band graph with Ag nanoparticles in ZnO film. Without Ag nanoparticles (top) or in darkness (middle), the hole transfer from the ITO film to ZnO is blocked by the energy level shift. Under the lighting (bottom), holes enter the ZnO film [61]. Copyright 2020, Nature Electronics.

2.4.3. MIR-to-Visible Upconversion

In 2020, Motmaen reported a novel and high-performance integrated chip which can realize the upconversion process from MIR (3–5 μm) to the green light (523 nm). This upconversion device was fabricated using a doped PbSe layer as a PD, an NPN (*n*-type, *p*-type, *n*-type) bipolar junction transistor (NPN-BJT) as a current amplifier, and an OLED as an emission layer; its structure is shown in Figure 9a, the energy band diagram is shown in Figure 9b. In this new type of upconverter, PbSe QDs were used to realize the photodetection and generate current, then the output electric information can be amplified by NPN-BJT. With the amplified current the LED can be driven and emits visible green light [62]. Under 3 μm MIR light illumination (0.5 mW/cm^2) (Figure 9c) and total voltage of 13.5 V, the EQE is calculated to 600% (Figure 9d), such high EQE is mainly thanks to the amplifier function of NPN-BJT.

The liquid semiconductor QDs synthesized by colloidal chemistry can precisely tune the bandgap by adjusting the size, and then achieve tunable infrared photoresponse in a wide spectral range. Therefore, from the above demonstration, it can be seen that the application of CQDs in the field of infrared-to-visible upconversion can realize infrared detection in all main infrared bands of NIR, SWIR, and MIR.

Therefore, after several decades of development, infrared-to-visible upconversion devices have shown a transition trend from traditional all-inorganic, all-organic, and organic–inorganic materials to emerging liquid CQD materials. The development of infrared-to-visible light upconversion devices based on CQDs has become one of the important trends to achieve wide detecting spectrum, high-performance, low-cost, flexible, and large-area infrared imaging in the future. The performance of the upconversion devices in the references discussed above is shown in Table 1.

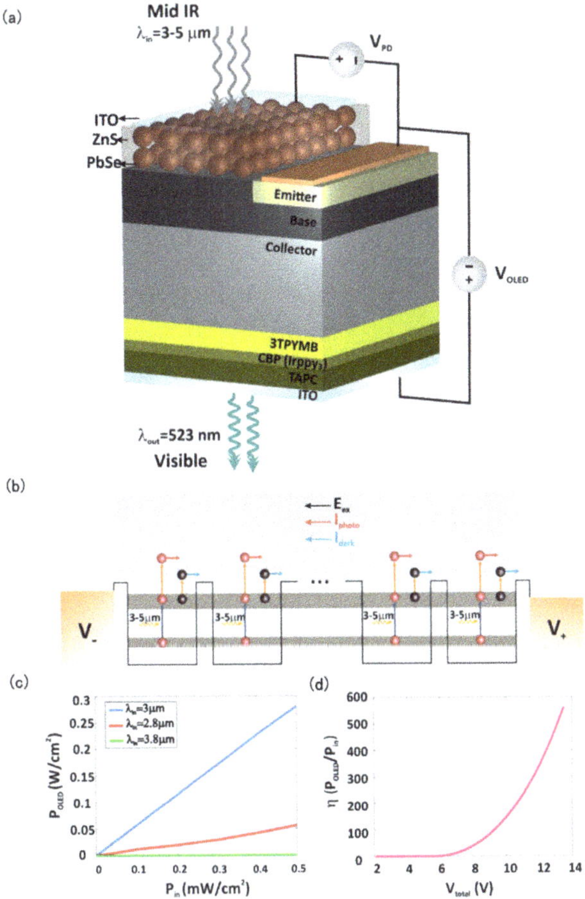

Figure 9. Structures and performance characterization of CQDs-based MIR-to-visible upconversion devices. (**a**) Structure of MIR-to-visible light upconversion device. (**b**) Schematic shows the energy band diagram of the PbSe/ ZnS quantum dots PD and the transition inside PD. The transfer of electrons from the energy levels of the excited state to the continuum by means of the thermal energy of the environment causes a dark current (black circle, I_{dark}). The photons excite electrons from energy levels of ground state to energy levels of excited state, resulting in a photocurrent (red circle, I_{photo}). (**c**) The output power density depending on the lighting power density MIR with different wavelengths. (**d**) The efficiency of PbSe upconversion device as function of the applied voltage under MIR light illumination (λ = 3 μm, 0.5 mW/cm^2) [62]. Copyright 2020, Scientific Reports.

Table 1. Progress in upconversion devices.

Year	Detect Material	Emit Material	Type	Detect Range (nm)	Emission	Maximum Brightness (cd/m^2)	Conversion Efficiency (%)	Ref.
1995	GaAs/AlGaAs [1]	InGaAs/GaAs	AI	1500	NIR (927 nm)	-	0.8	[29]
2000	InGaAsP	InAsP	AI	~1600	NIR (1000 nm)	-	-	[30]
2004	InGaAs/InP	GaAs/AlGaAs	AI	~1500	NIR (870 nm)	-	76	[31]
2009	GaNAsSb/GaAs	GaAs/AlGaAs	AI	1300	NIR (950 nm)	-	4.8	[32]
2007	PVK:TNFDM [2]	Alq$_3$	AO	810	Green (530 nm)	-	-	[33]
2010	SnPc:C$_{60}$	CBP: Ir(ppy)$_3$	AO	~830	Green	853	2.7	[34]
2015	ClAlPc:C70	CBP: Ir(ppy)$_3$	AO	~780	Green	1553	-	[36]
2017	ING-T-DPP: PC61BM	4CzIPN:2CzPN :4CzTPN-Ph	AO	~810	Full-color	150	0.11	[37]
2018	PDPP3T-PCBM	Be(pp)$_2$: Ir(ppy)$_2$(acac)	AO	~850	Green	1504	29.6	[35]
2018	PbPc:C$_{60}$	2PXZ-OXD	AO	808~900	Green	32935	256	[38]
2018	SQ-880: PCBM	Alq$_3$	AO	~1000	Green	313	~0.25–0.5	[41]
2019	SQ-880: PCBM	SY	AO	~980	Yellow	760	1.6	[39]
2019	PBDTT-BTQ	CBP: Ir(ppy)$_3$	AO	700~1100	Green (520 nm)	<10	~0.8	[40]
2020	DPP–DTT: SWIR dye	CsPbBr$_3$	AO	1000~1400 (peak:1050)	Green (516 nm)	-	>0.1	[42]
2007	InGaAs/InP	Alq:C545T	HOI	1500	Green (520 nm)	1520	0.7(W/W)	[49]
2007	InGaAs/InP	Alq:C545T	HOI	1500	Green (520 nm)	~600	-	[50]
2008	InGaAs/InP	Alq:C545T	HOI	~1500	Green (543 nm)	1580	~1.5(W/W)	[43]
2010	InGaAs/InP	Alq$_3$:C545T	HOI	~1500	Green (564 nm)	~9000	59	[44]
2012	InGaAs/InP	Alq$_3$	HOI	~1500	Green (520 nm)	200	0.2	[45]
2018	FAPbI$_3$	CBP: Ir(ppy)$_2$(acac)	HOI	790~900 (peak:830)	Green (520 nm)	3.3	-	[46]
2018	GaAs	Ga$_{0.5}$In$_{0.5}$P: Al$_{0.15}$Ga$_{0.35}$In$_{0.5}$P	HOI	~810	Red (630 nm) Yellow (590 nm)	-	1.5	[47]
2019	GaAs	AlGaInP	HOI	~810	Red (630 nm)	-	-	[48]
2011	PbSe	CBP: Ir(ppy)$_3$	CQD	1500 (peak:1300)	Green (550 nm)	5	1.3	[57]
2016	PbS	TATC: Ir(ppy)$_3$	CQD	940~1042	Green (550 nm)	~100	1597	[58]
2018	DPP–DTT: COi8DFIC	CsPbBr$_3$	CQD	850	Green (520 nm)	~300	1.7	[59]
2020	PbS	CdSe	CQD	970	Red (624 nm)	155	6.3	[60]
2020	PbS	CdSe/ZnS	CQD	400~1600	Green (525 nm)	~600	6.5	[61]
2020	PbSe	CBP: Ir(ppy)$_3$	CQD	3000~5000	Green (523 nm)	-	-	[62]

[1] Materials are mixed together; [2] Materials are integrated but not mixed together; Abbreviations: AI: all-inorganic; AO: all-organic; HOI: hybrid organic–inorganic; Alq$_3$, Alq: tris-8-hydroxyquinolineAluminum; SnPc: tin phthalocyanine; CBP: 4,4-N, N-dicarbazole-biphenyl; Ir(ppy)$_3$: fac-tris(2-phenylpyridinato) iridium (III); PDPP3T: Poly[{2,5-bis(2-hexyldecyl)-2,3,5,6-tetrahydro-3,6-diox-opyrrolo[3,4-c]pyrrole-1,4-diyl} -alt-{[2,2′:5′,2″-terthiophene]-5,5″-diyl}]; PCBM: [6,6]-phenyl-C61-butyric acid methyl ester; Be(pp)$_2$: Bis[2-(2-hydroxyphenyl)-pyridine] beryllium; Ir(ppy)$_2$(acac): bis(2-phenylpyridine) (acetylacetonato) iridium (III); ClAlPc: chloroaluminum phthalocyanine; PC$_{61}$BM: [6,6]- phenyl-C61-butyric acid methyl ester; 4CzIPN: 1,2,3,5-tetrakis(carbazol-9-yl)-4,6-dicyanobenzene; 2CzPN: 1,2-bis(carbazol-9-yl)-4,5-dicyanobenzene; 4CzTPN-Ph: 1,4-dicyano-2,3,5,6-tetrakis (3,6-diphenylcarbazol-9-yl)benzene; 2PXZ-OXD: 2,5-bis(4-(10-phenoxazyl)phenyl)-1,3,4-oxadiazole; SY: fluorescent poly(para-phenylene vinylene) copolymer called super yellow; PBDTT-BTQ: poly 4-(4,8-bis(5-(2-butyloctyl)-thiophen-2-yl)benzo[1,2-b:4,5-b′]dithiophen-2-yl) -6,7-diethyl-[1,2,5] thiadiazolo [3,4-g]quinoxaline; C545T: 10-2-benzothiazoly-1,1,7,7–tetramethyl-2,3,6,7-tetrahydro-1H, 5H, 11H-[1] benzopyrano [6,7,8-ij] quinolizine-11-one.

3. Challenges and Perspectives

Although infrared upconversion devices have made many remarkable signs of progress in the past few years and some advantages of them are irreplaceable, there are still some important problems that need further consideration.

1. The detecting wavelength of infrared-to-visible upconverters is mostly concentrated in the near-infrared range. Although some SWIR and MIR spectral range devices are reported, those upconversion devices still have to improve photon-to-photon conversion efficiency, reduce the turn-on voltage and optimize other performance parameters. The short detection wavelength of upconversion devices limits their application scenarios.
2. The upconversion efficiency and sensitivity of infrared-to-visible light upconversion devices are relatively low. The upconversion efficiency of the currently reported devices is difficult to exceed 10%, leading to most of the existing devices only operating under intense infrared light (usually an infrared incident power range of 10–100 mW cm^{-2}).
3. In many medical and commercial fields, the performance of flexible devices is very important. Even though some all-solution-based processes can be achieved, with these methods the flexible devices' performance characterization is scarce, meanwhile, the brightness and life time of flexible devices are unsatisfactory.

Overall, the performance of infrared-to-visible upconversion devices remains to be significantly enhanced to meet application requirements. High conversion efficiency, low turn-on voltage, flexible device, and wide detection band are vital to upconverters. To meet these above challenges, we need to pay more attention to the following aspects.

1. Low dimensional materials, such as zero-dimensional materials (CQD), one-dimensional materials (III-Sb nanowires), and two-dimensional materials (quantum well, graphene, and black phosphorus) possess tunable bandgaps, high carrier mobility, precisely controlled materials growth process, low cost, and solution process compatibility, thus they have been widely applied in both infrared detectors and LED fields. In the future, there may be a strong trend of low dimensional materials replacing traditional materials to be utilized in infrared-to-visible upconversion devices.
2. The upconversion efficiency and sensitivity of the infrared-to-visible upconversion device are the keys to determining whether it can detect weak infrared light in nature for practical use. Therefore, the high upconversion efficiency and sensitivity to meet the practical application of upconversion infrared imaging may become important trends in future development.
3. To manufacture a flexible device with stable performance and to characterize it accordingly, it is necessary to optimize and improve the existing manufacturing materials and manufacturing methods of the upconverter. The current flexible devices are based on CQD materials, but these materials are often very toxic and cannot be used in large-scale production applications. Therefore, only flexible devices with high performance and low toxicity can be used in biology and other fields.
4. Broaden the wavelength range of upconversion. In fact, due to the invisible light range from 700 nm to 1100 nm for the naked eye, it can be easily detected by traditional silicon PDs. Thus, it could be said that the benefits of the upconversion device concept only become obvious when NIR light with wavelengths beyond the silicon band can be upconverted into visible light. The current upconversion devices are mainly concentrated in the NIR band, and the research on SWIR and above infrared upconversion devices is much less. Although CQD semiconductor materials have the ability to cover the main wavelength bands of infrared, the maximum infrared detection wavelength (1.6 μm) of all-CQDs-based infrared-to-visible upconversion devices, which just reaches the SWIR and limits its application field. Therefore, it is a mainstream direction for future research and development to break through the infrared detection band range of infrared-to-visible upconversion devices, thereby broadening their application field.

Author Contributions: Conceptualization, X.T., T.R. and G.M.; methodology, M.C.; software, X.T., T.R. and G.M.; validation, X.T., T.R. and G.M.; formal analysis, M.C.; investigation, X.T., T.R. and G.M.; resources, X.T., T.R. and G.M.; data curation, X.T., T.R. and G.M.; writing—original draft preparation, T.R., G.M.; writing—review and editing, X.T.; visualization, X.T., T.R. and G.M.; supervision, X.T.; project administration, X.T.; funding acquisition, X.T. All authors have read and agreed to the published version of the manuscript.

Funding: This research was funded by the National Natural Science Foundation of China and the National Key R&D Program of China (NSFC No. 62035004, 2021YFA0717600, and NSFC No. 62105022).

Institutional Review Board Statement: Not applicable.

Informed Consent Statement: Not applicable.

Data Availability Statement: Not applicable.

Conflicts of Interest: The authors declare no conflict of interest.

References

1. Minotto, A.; Haigh, P.A.; Łukasiewicz, Ł.G.; Lunedei, E.; Gryko, D.T.; Darwazeh, I.; Cacialli, F. Visible light communication with efficient far-red/near-infrared polymer light-emitting diodes. *Light Sci. Appl.* **2020**, *9*, 70. [CrossRef] [PubMed]
2. Kim, S.; Lim, Y.T.; Soltesz, E.G.; De Grand, A.M.; Lee, J.; Nakayama, A.; Parker, J.A.; Mihaljevic, T.; Laurence, R.G.; Dor, D.M.; et al. Near-infrared fluorescent type II quantum dots for sentinel lymph node mapping. *Nat. Biotechnol.* **2004**, *22*, 93–97. [CrossRef] [PubMed]
3. Schreiner, K. Night Vision: Infrared takes to the road. *IEEE Comput. Graph. Appl.* **1999**, *19*, 6–10. [CrossRef]
4. Bellotti, C.; Bellotti, F.; De Gloria, A.; Andreone, L.; Mariani, M. Developing a near infrared based night vision system. *IEEE Intell. Veh. Symp. Proc.* **2004**, *2004*, 686–691. [CrossRef]
5. Chorier, P.; Tribolet, P.M.; Fillon, P.; Manissadjian, A. Application needs and trade-offs for short-wave infrared detectors. *Infrared Technol. Appl. XXIX* **2003**, *5074*, 363. [CrossRef]
6. Nelson, M.P.; Shi, L.; Zbur, L.; Priore, R.J.; Treado, P.J. Real-time short-wave infrared hyperspectral conformal imaging sensor for the detection of threat materials. *Chem. Biol. Radiol. Nucl. Explos. Sens. XVII* **2016**, *9824*, 982416. [CrossRef]
7. Wang, M.; Knobelspiesse, K.D.; McClain, C.R. Study of the Sea-Viewing Wide Field-of-View Sensor (SeaWiFS) aerosol optical property data over ocean in combination with the ocean color products. *J. Geophys. Res. D Atmos.* **2005**, *110*, D10S06. [CrossRef]
8. Keuleyan, S.; Lhuillier, E.; Brajuskovic, V.; Guyot-Sionnest, P. Mid-infrared HgTe colloidal quantum dot photodetectors. *Nat. Photonics* **2011**, *5*, 489–493. [CrossRef]
9. Ackerman, M.M.; Tang, X.; Guyot-Sionnest, P. Fast and Sensitive Colloidal Quantum Dot Mid-Wave Infrared Photodetectors. *ACS Nano* **2018**, *12*, 7264–7271. [CrossRef] [PubMed]
10. Ackerman, M.M.; Chen, M.; Guyot-Sionnest, P. HgTe colloidal quantum dot photodiodes for extended short-wave infrared detection. *Appl. Phys. Lett.* **2020**, *116*, 083502. [CrossRef]
11. Amani, M.; Regan, E.; Bullock, J.; Ahn, G.H.; Javey, A. Mid-Wave Infrared Photoconductors Based on Black Phosphorus-Arsenic Alloys. *ACS Nano* **2017**, *11*, 11724–11731. [CrossRef] [PubMed]
12. Hoang, A.M.; Dehzangi, A.; Adhikary, S.; Razeghi, M. High performance bias-selectable three-color Short-wave/Mid-wave/Long-wave Infrared Photodetectors based on Type-II InAs/GaSb/AlSb superlattices. *Sci. Rep.* **2016**, *6*, 24144. [CrossRef] [PubMed]
13. Rogalski, A. HgCdTe infrared detector material: History, status and outlook. *Rep. Prog. Phys.* **2005**, *68*, 2267–2336. [CrossRef]
14. Nienhaus, L.; Correa-Baena, J.-P.; Wieghold, S.; Einzinger, M.; Lin, T.-A.; Shulenberger, K.E.; Klein, N.D.; Wu, M.; Bulović, V.; Buonassisi, T.; et al. Triplet-Sensitization by Lead Halide Perovskite Thin Films for Near-Infrared-to-Visible Upconversion. *ACS Energy Lett.* **2019**, *4*, 888–895. [CrossRef]
15. Wu, M.; Lin, T.-A.; Tiepelt, J.O.; Bulović, V.; Baldo, M.A. Nanocrystal-Sensitized Infrared-to-Visible Upconversion in a Microcavity under Subsolar Flux. *Nano Lett.* **2021**, *21*, 1011–1016. [CrossRef] [PubMed]
16. Lissau, J.S.; Khelfallah, M.; Madsen, M. Near-Infrared to Visible Photon Upconversion by Palladium(II) Octabutoxyphthalocyanine and Rubrene in the Solid State. *J. Phys. Chem. C* **2021**, *125*, 25643–25650. [CrossRef]
17. Xu, Z.; Huang, Z.; Li, C.; Huang, T.; Evangelista, F.A.; Tang, M.L.; Lian, T. Tuning the Quantum Dot (QD)/Mediator Interface for Optimal Efficiency of QD-Sensitized Near-Infrared-to-Visible Photon Upconversion Systems. *ACS Appl. Mater. Interfaces* **2020**, *12*, 36558–36567. [CrossRef]
18. VanOrman, Z.A.; Nienhaus, L. Bulk Metal Halide Perovskites as Triplet Sensitizers: Taking Charge of Upconversion. *ACS Energy Lett.* **2021**, *6*, 3686–3694. [CrossRef]
19. Bharmoria, P.; Bildirir, H.; Moth-Poulsen, K. Triplet–triplet annihilation based near infrared to visible molecular photon upconversion. *Chem. Soc. Rev.* **2020**, *49*, 6529–6554. [CrossRef]
20. Wu, M.; Congreve, D.N.; Wilson, M.W.B.; Jean, J.; Geva, N.; Welborn, M.; Van Voorhis, T.; Bulović, V.; Bawendi, M.G.; Baldo, M.A. Solid-state infrared-to-visible upconversion sensitized by colloidal nanocrystals. *Nat. Photonics* **2016**, *10*, 31–34. [CrossRef]

21. Ye, C.; Zhou, L.; Wang, X.; Liang, Z. Photon upconversion: From two-photon absorption (TPA) to triplet–triplet annihilation (TTA). *Phys. Chem. Chem. Phys.* **2016**, *18*, 10818–10835. [CrossRef]
22. Huang, Z.; Xu, Z.; Mahboub, M.; Liang, Z.; Jaimes, P.; Xia, P.; Graham, K.R.; Tang, M.L.; Lian, T. Enhanced Near-Infrared-to-Visible Upconversion by Synthetic Control of PbS Nanocrystal Triplet Photosensitizers. *J. Am. Chem. Soc.* **2019**, *141*, 9769–9772. [CrossRef] [PubMed]
23. Liu, Y.; Lu, Y.; Yang, X.; Zheng, X.; Wen, S.; Wang, F.; Vidal, X.; Zhao, J.; Liu, D.; Zhou, Z.; et al. Amplified stimulated emission in upconversion nanoparticles for super-resolution nanoscopy. *Nature* **2017**, *543*, 229–233. [CrossRef] [PubMed]
24. Wen, S.; Zhou, J.; Zheng, K.; Bednarkiewicz, A.; Liu, X.; Jin, D. Advances in highly doped upconversion nanoparticles. *Nat. Commun.* **2018**, *9*, 2415. [CrossRef] [PubMed]
25. Tao, Y.; Huang, A.J.Y.; Hashimotodani, Y.; Kano, M.; All, A.H.; Tsutsui-kimura, I.; Tanaka, K.F.; Liu, X.; Mchugh, T.J. NIR deep brain stimulation via upconversion nanoparticle-mediated optogenetics. *Science* **2018**, *684*, 679–684.
26. Xu, J.; Xu, L.; Wang, C.; Yang, R.; Zhuang, Q.; Han, X.; Dong, Z.; Zhu, W.; Peng, R.; Liu, Z. Near-Infrared-Triggered Photodynamic Therapy with Multitasking Upconversion Nanoparticles in Combination with Checkpoint Blockade for Immunotherapy of Colorectal Cancer. *ACS Nano* **2017**, *11*, 4463–4474. [CrossRef] [PubMed]
27. Mita, Y. Detection of 1.5-μm wavelength laser light emission by infrared-excitable phosphors. *Appl. Phys. Lett.* **1981**, *39*, 587–589. [CrossRef]
28. Wang, Y.; Ohwaki, J. High-efficiency infrared-to-visible upconversion of Er^{3+} in $BaCl_2$. *J. Appl. Phys.* **1993**, *74*, 1272–1278. [CrossRef]
29. Liu, H.C.; Li, J.; Wasilewski, Z.R.; Buchanan, M. Integrated quantum well intersub-band photodetectorand light emitting diode. *Electron. Lett.* **1995**, *31*, 832–833. [CrossRef]
30. Liu, H.C.; Gao, M.; Poole, P.J. 1.5 μm up-conversion device. *Electron. Lett.* **2000**, *36*, 1300–1301. [CrossRef]
31. Ban, D.; Luo, H.; Liu, H.C.; Wasilewski, Z.R.; Springthorpe, A.J.; Glew, R.; Buchanan, M. Optimized GaAs/AlGaAs light-emitting diodes and high efficiency wafer-fused optical up-conversion devices. *J. Appl. Phys.* **2004**, *96*, 5243–5248. [CrossRef]
32. Yang, Y.; Shen, W.Z.; Liu, H.C.; Laframboise, S.R.; Wicaksono, S.; Yoon, S.F.; Tan, K.H. Near-infrared photon upconversion devices based on GaNAsSb active layer lattice matched to GaAs. *Appl. Phys. Lett.* **2009**, *94*, 2007–2010. [CrossRef]
33. Lu, J.; Zheng, Y.; Chen, Z.; Xiao, L.; Gong, Q. Optical upconversion devices based on photosensitizer-doped organic light-emitting diodes. *Appl. Phys. Lett.* **2007**, *91*, 201107. [CrossRef]
34. Kim, D.Y.; Song, D.W.; Chopra, N.; De Somer, P.; So, F. Organic infrared upconversion device. *Adv. Mater.* **2010**, *22*, 2260–2263. [CrossRef] [PubMed]
35. Yang, D.; Zhou, X.; Ma, D.; Vadim, A.; Ahamad, T.; Alshehri, S.M. Near infrared to visible light organic up-conversion devices with photon-to-photon conversion efficiency approaching 30%. *Mater. Horiz.* **2018**, *5*, 874–882. [CrossRef]
36. Liu, S.W.; Lee, C.C.; Yuan, C.H.; Su, W.C.; Lin, S.Y.; Chang, W.C.; Huang, B.Y.; Lin, C.F.; Lee, Y.Z.; Su, T.H.; et al. Transparent organic upconversion devices for near-infrared sensing. *Adv. Mater.* **2015**, *27*, 1217–1222. [CrossRef] [PubMed]
37. Tachibana, H.; Aizawa, N.; Hidaka, Y.; Yasuda, T. Tunable Full-Color Electroluminescence from All-Organic Optical Upconversion Devices by Near-Infrared Sensing. *ACS Photonics* **2017**, *4*, 223–227. [CrossRef]
38. Song, Q.; Lin, T.; Su, Z.; Chu, B.; Yang, H.; Li, W.; Lee, C.S. Organic upconversion display with an over 100% photon-to-photon upconversion efficiency and a simple pixelless device structure. *J. Phys. Chem. Lett.* **2018**, *9*, 6818–6824. [CrossRef]
39. Strassel, K.; Ramanandan, S.P.; Abdolhosseinzadeh, S.; Diethelm, M.; Nuesch, F.; Hany, R. Solution-processed organic optical upconversion device. *ACS Appl. Mater. Interfaces* **2019**, *11*, 23428–23435. [CrossRef]
40. Yeddu, V.; Seo, G.; Cruciani, F.; Beaujuge, P.M.; Kim, D.Y. Low-Band-Gap Polymer-Based Infrared-to-Visible Upconversion Organic Light-Emitting Diodes with Infrared Sensitivity up to 1.1 μm. *ACS Photonics* **2019**, *6*, 2368–2374. [CrossRef]
41. Strassel, K.; Kaiser, A.; Jenatsch, S.; Véron, A.C.; Anantharaman, S.B.; Hack, E.; Diethelm, M.; Nüesch, F.; Aderne, R.; Legnani, C.; et al. Squaraine Dye for a Visibly Transparent All-Organic Optical Upconversion Device with Sensitivity at 1000 nm. *ACS Appl. Mater. Interfaces* **2018**, *10*, 11063–11069. [CrossRef] [PubMed]
42. Li, N.; Lan, Z.; Lau, Y.S.; Xie, J.; Zhao, D.; Zhu, F. SWIR Photodetection and Visualization Realized by Incorporating an Organic SWIR Sensitive Bulk Heterojunction. *Adv. Sci.* **2020**, *7*, 2000444. [CrossRef] [PubMed]
43. Chen, J.; Ban, D.; Feng, X.; Lu, Z.; Fathololoumi, S.; Springthorpe, A.J.; Liu, H.C. Enhanced efficiency in near-infrared inorganic/organic hybrid optical upconverter with an embedded mirror. *J. Appl. Phys.* **2008**, *103*, 103112. [CrossRef]
44. Chen, J.; Ban, D.; Helander, M.G.; Lu, Z.H.; Poole, P. Near-infrared inorganic/organic optical upconverter with an external power efficiency of >100%. *Adv. Mater.* **2010**, *22*, 4900–4904. [CrossRef] [PubMed]
45. Chen, J.; Tao, J.; Ban, D.; Helander, M.G.; Wang, Z.; Qiu, J.; Lu, Z. Hybrid organic/inorganic optical up-converter for pixel-less near-infrared imaging. *Adv. Mater.* **2012**, *24*, 3138–3142. [CrossRef]
46. Yu, B.H.; Cheng, Y.; Li, M.; Tsang, S.W.; So, F. Sub-Band Gap Turn-On Near-Infrared-to-Visible Up-Conversion Device Enabled by an Organic-Inorganic Hybrid Perovskite Photovoltaic Absorber. *ACS Appl. Mater. Interfaces* **2018**, *10*, 15920–15925. [CrossRef] [PubMed]
47. Ding, H.; Lu, L.; Shi, Z.; Wang, D.; Li, L.; Li, X.; Ren, Y.; Liu, C.; Cheng, D.; Kim, H.; et al. Microscale optoelectronic infrared-to-visible upconversion devices and their use as injectable light sources. *Proc. Natl. Acad. Sci. USA* **2018**, *115*, 6632–6637. [CrossRef] [PubMed]

48. Shi, Z.; Ding, H.; Hong, H.; Cheng, D.; Rajabi, K.; Yang, J.; Wang, Y.; Wang, L.; Luo, Y.; Liu, K.; et al. Ultrafast and low-power optoelectronic infrared-to-visible upconversion devices. *Photonics Res.* **2019**, *7*, 1161. [CrossRef]
49. Ban, D.; Han, S.; Lu, Z.H.; Oogarah, T.; Springthorpe, A.J.; Liu, H.C. Near-infrared to visible light optical upconversion by direct tandem integration of organic light-emitting diode and inorganic photodetector. *Appl. Phys. Lett.* **2007**, *90*, 093108. [CrossRef]
50. Ban, D.; Han, S.; Lu, Z.H.; Oogarah, T.; Springthorpe, A.J.; Liu, H.C. Organic—Inorganic Hybrid Optical Upconverter. *IEEE Trans. Electron Devices* **2007**, *54*, 1645–1650. [CrossRef]
51. Keuleyan, S.; Lhuillier, E.; Guyot-Sionnest, P. Synthesis of colloidal HgTe quantum dots for narrow mid-IR emission and detection. *J. Am. Chem. Soc.* **2011**, *133*, 16422–16424. [CrossRef] [PubMed]
52. Tang, X.; Ackerman, M.M.; Guyot-Sionnest, P. Thermal Imaging with Plasmon Resonance Enhanced HgTe Colloidal Quantum Dot Photovoltaic Devices. *ACS Nano* **2018**, *12*, 7362–7370. [CrossRef] [PubMed]
53. Chen, M.; Lan, X.; Tang, X.; Wang, Y.; Hudson, M.H.; Talapin, D.V.; Guyot-Sionnest, P. High Carrier Mobility in HgTe Quantum Dot Solids Improves Mid-IR Photodetectors. *ACS Photonics* **2019**, *6*, 2358–2365. [CrossRef]
54. Tang, X.; Ackerman, M.M.; Chen, M.; Guyot-Sionnest, P. Dual-band infrared imaging using stacked colloidal quantum dot photodiodes. *Nat. Photonics* **2019**, *13*, 277–282. [CrossRef]
55. Tang, X.; Chen, M.; Kamath, A.; Ackerman, M.M.; Guyot-Sionnest, P. Colloidal Quantum-Dots/Graphene/Silicon Dual-Channel Detection of Visible Light and Short-Wave Infrared. *ACS Photonics* **2020**, *7*, 1117–1121. [CrossRef]
56. Zhang, S.; Chen, M.; Mu, G.; Li, J.; Hao, Q.; Tang, X. Spray-Stencil Lithography Enabled Large-Scale Fabrication of Multispectral Colloidal Quantum-Dot Infrared Detectors. *Adv. Mater. Technol.* **2021**, *2101132*, 1–8. [CrossRef]
57. Kim, D.Y.; Choudhury, K.R.; Lee, J.W.; Song, D.W.; Sarasqueta, G.; So, F. PbSe Nanocrystal-Based Infrared-to-Visible Up-Conversion Device. *Nano Lett.* **2011**, *11*, 2109–2113. [CrossRef] [PubMed]
58. Yu, H.; Kim, D.; Lee, J.; Baek, S.; Lee, J.; Singh, R.; So, F. High-gain infrared-to-visible upconversion light-emitting phototransistors. *Nat. Photonics* **2016**, *10*, 129–134. [CrossRef]
59. Li, N.; Lau, Y.S.; Xiao, Z.; Ding, L.; Zhu, F. NIR to Visible Light Upconversion Devices Comprising an NIR Charge Generation Layer and a Perovskite Emitter. *Adv. Opt. Mater.* **2018**, *6*, 1801084. [CrossRef]
60. Tang, H.; Shi, K.; Zhang, N.; Wen, Z.; Xiao, X.; Xu, B.; Butt, H.; Pikramenou, Z.; Wang, K.; Sun, X.W. Up-conversion device based on quantum dots with high-conversion efficiency over 6%. *IEEE Access* **2020**, *8*, 71041–71049. [CrossRef]
61. Zhou, W.; Shang, Y.; García de Arquer, F.P.; Xu, K.; Wang, R.; Luo, S.; Xiao, X.; Zhou, X.; Huang, R.; Sargent, E.H.; et al. Solution-processed upconversion photodetectors based on quantum dots. *Nat. Electron.* **2020**, *3*, 251–258. [CrossRef]
62. Motmaen, A.; Rostami, A.; Matloub, S. Ultra High-efficiency Integrated Mid Infrared to Visible Up-conversion System. *Sci. Rep.* **2020**, *10*, 9325. [CrossRef] [PubMed]

Communication

Simulation of Monolithically Integrated Meta-Lens with Colloidal Quantum Dot Infrared Detectors for Enhanced Absorption

Yan Ning, Shuo Zhang, Yao Hu, Qun Hao * and Xin Tang *

Department of Optoelectronics, Beijing Institute of Technology, Beijing 100081, China; ningyan516@163.com (Y.N.); 3120195342@bit.edu.cn (S.Z.); huy08@bit.edu.cn (Y.H.)
* Correspondence: qhao@bit.edu.cn (Q.H.); xintang@bit.edu.cn (X.T.)

Received: 13 October 2020; Accepted: 4 December 2020; Published: 14 December 2020

Abstract: Colloidal quantum dots (CQDs) have been intensively investigated over the past decades in various fields for both light detection and emission applications due to their advantages like low cost, large-scale production, and tunable spectral absorption. However, current infrared CQD detectors still suffer from one common problem, which is the low absorption rate limited by CQD film thickness. Here, we report a simulation study of CQD infrared detectors with monolithically integrated meta-lenses as light concentrators. The design of the meta-lens for 4 μm infrared was investigated and simulation results show that light intensity in the focused region is ~20 times higher. Full device stacks were also simulated, and results show that, with a meta-lens, high absorption of 80% can be achieved even when the electric area of the CQD detectors was decreased by a factor of 64. With higher absorption and a smaller detector area, the employment of meta-lenses as optical concentrators could possibly improve the detectivity by a factor of 32. Therefore, we believe that integration of CQD infrared detectors with meta-lenses could serve as a promising route towards high performance infrared optoelectronics.

Keywords: colloidal quantum dots; meta-lens; detectivity; optical concentrator

---

1. Introduction

With steady advancement, the maximum absorption wavelength of colloidal quantum dots (CQDs) has been extended from near-infrared to longer wavelength, providing a promising alternative to bulk infrared semiconductors like HgCdTe, InSb and type-II superlattices [1,2]. Among all the colloidal nanomaterials, mercury telluride (HgTe) CQDs have so far demonstrated the highest spectral tunability from the short-wave infrared (SWIR) [3,4] and mid-wave infrared (MWIR) [5–9] to the long-wave infrared (LWIR) [10,11] and terahertz (THz) region [12]. HgTe CQD-based photoconductors [7,8,13], photovoltaics [5,6,9], and phototransistors [14,15] have all been studied. Among them, HgTe CQD photovoltaic detectors have unique advantages like the absence of 1/*f* noise [9] and fast operation speed [5]. SWIR detectors with detectivity up to 5×10^{10} Jones at room temperature [16], MWIR detectors with background-limited infrared photodetection (BLIP) performance at cryogenic temperature [6], SWIR hyperspectral detectors with 64 channels [17], CQD-silicon dual-channel detectors [18], and SWIR/MWIR dual-band imagers with vertical junctions [19] have all been demonstrated in photovoltaic configuration. Despite all the exciting achievements, most infrared CQD detectors still suffer from one common problem, which is the low absorption rate limited by the film thickness of CQDs. CQD detectors are usually made by layer-by-layer drop-casting, and for an infrared HgTe CQD detector, the regular thickness is ~400 nm and only 20% of incident light can be absorbed in the MWIR region [6]. Further increase of film thickness would lead to cracks and delamination due to mechanical stress from the drying and ligand exchange process [20]. Therefore, to improve the performance of infrared CQD photodetectors, one needs to find a strategy to enhance light absorption without increasing CQD film thickness.

For infrared detectors, the key figure of merit is detectivity, D^*, which can be expressed by

$$D^* = \frac{\sqrt{A}}{I_n}\Re \tag{1}$$

where A is the light sensing area, I_n is noise current, and $\Re$ is responsivity, which is related to absorption α by

$$\Re = \frac{\alpha e}{hv}\eta \tag{2}$$

where e is elementary charge, h is the Planck constant, v is the frequency of photon, and η is internal quantum efficiency. Therefore, the absorption of CQD film is related to D^* by

$$D^* = \frac{\sqrt{A}}{I_n}\frac{\alpha e}{hv}\eta \tag{3}$$

Strategies of using metallic plasmonic nanostructures, resonant cavities, and photonic crystals have all been reported to boost CQD absorption up to 2–3 times [6,16]. Those structures are fabricated using regular lithographic processes [11] or multi-beam laser interference lithography [21,22], which may change the surface condition of CQDs and introduce noise source into CQD detectors. To optimize D^*, it is highly favorable to find a way to increase light absorption and decrease noise at the same time. The noise of a CQD photovoltaic detector mainly consists of Johnson noise, which scales with detector electric area A_e by a factor of $A_e^{1/2}$ [18]. Employment of optical concentrators, therefore, seems to be a suitable method to address such a challenge. With light concentrators, the electric area of a detector can be decreased while maintaining the same responsivity by improving light absorption, leading to enhanced detectivity. Compared to three-dimensional (3D) microlenses or microspheres, meta-lenses can be fabricated in scalable ways and can provide functionalities like controlling the phase, amplitude, and polarization of the incident light by simply altering the shape and the size of each meta-unit [23,24]. HgCdTe and GaSb detectors with meta-lenses have been reported. However, such a concept has rarely been investigated for infrared CQD photodetectors.

Here, we report a simulation study of CQD infrared detectors with a monolithically integrated meta-lens as light concentrator. The target spectral range is MWIR (3–5 μm). The design of the meta-lens in the center of MWIR (4 μm) is discussed. The simulation results showed that light intensity in the focused region is ~20 times higher than that of the incident light. More importantly, we find that the chromatic aberration of the meta-lens might unintentionally help the development of spectrally selective infrared detectors, which can be used to distinguish wavelength information of incident light. Besides simulation of the meta-lens, full device stacks were also simulated, and results showed that with a meta-lens, absorption of 80% can be achieved even when the electric area of the CQD detector was decreased by a factor of 64. Therefore, we believe that integration of CQD infrared detectors with a meta-lens could serve as a promising route towards high performance infrared optoelectronics.

2. Modeling and Simulation

COMSOL Multiphysics is used to conduct the simulation in our study. To save computation resources, two-dimensional (2D) simulation is used as shown in Figure 1. The light is introduced from the "incident port" with controlled power of 1 W. A listening port is placed on the top of the model. The left and right sides are set as "Scattering Boundary Condition" to avoid light reflection. The substrate material is chosen to be Al_2O_3 with a refractive index of $1.6 + 0i$. The meta-lens units are made of silicon with a refractive index of $3.5 + 0i$. For HgTe CQDs, the refractive index is $2.3 + 0.1i$ [6]. The rest of the model is set to be air with a refractive index of $1 + 0i$. The absorption of CQDs is calculated by comparing the incident power and outlet power. The wavelength of incident light is swept from 1.5 to 5.0 μm to calculate the spectral absorption. In out simulation, the diameter of the meta-lens is set to be 200 μm to match with our commonly used size of detector in experimental tests. The phase profile and meta-lens design are generated by using open source software called MetaOptics (version v2) [25]. In our design, we use an unpolarized plane wave with uniform intensity distribution across the meta-lens region. To lower fabrication complexity, a pillar structure with periodicity of 2 μm and height of 6 μm was used. Complete phase shift of 0–2π can be realized by changing the diameter from 0.6 to 1.7 μm. Further shrinking the size of the meta-lens (periodicity, diameter, and height) should still work as long as the light can still be focused onto the CQD detector.

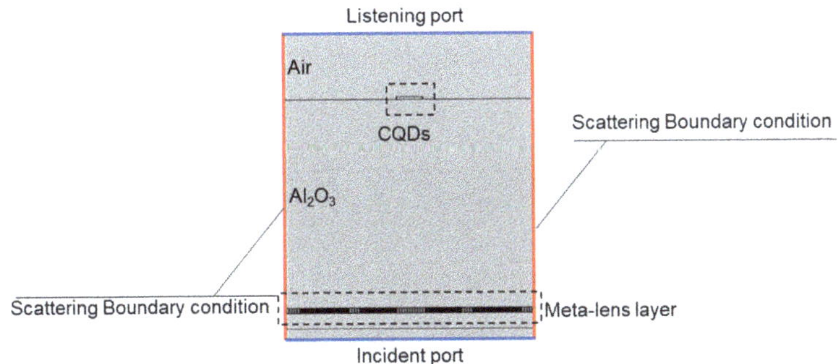

Figure 1. Simulation model of CQD detector integrated with a meta-lens.

3. Results and Discussions

The proposed design of CQD photodetectors with a monolithically integrated meta-lens is illustrated in Figure 2a. Pillar structures made of silicon were selected as the material to construct the meta-lens unit. The incident light illuminated on the meta-lens can be concentrated and focused into the CQD detector. As shown in the cross-sectional illustration in Figure 2b, the electric area of the CQD detector is much smaller than the optical area of the meta-lens. To design a functional meta-lens, the first step is to calculate the required phase profile based on the lens equation

$$\phi(x,y) = \frac{2\pi}{\lambda} n_m (f - \sqrt{x^2 + y^2 + f^2}) \tag{4}$$

where λ is the targeted wavelength, n_m is the index of refraction of the medium, and f is the focal length. For a regular imaging lens, the phase profile for λ = 4 μm and f = 300 μm is shown in Figure 2c (black curve). For a meta-lens, the continuous phase profile modulation is achieved by periodically changing the effective index of refraction of the medium, which can be done by tuning the height and diameter of the silicon pillar in the meta-lens unit. The red curve in Figure 2c shows the required phase profile of a meta-lens, based on which the meta-lens design was generated using open source software called MetaOptics. In our design, a pillar structure with periodicity of 2 μm and height of 6 μm was used. By changing the diameter from 0.6 to 1.7 μm, a phase shift of 0–2π was covered. Figure 2d shows the 2D phase profile mapping and structure design of a meta-lens.

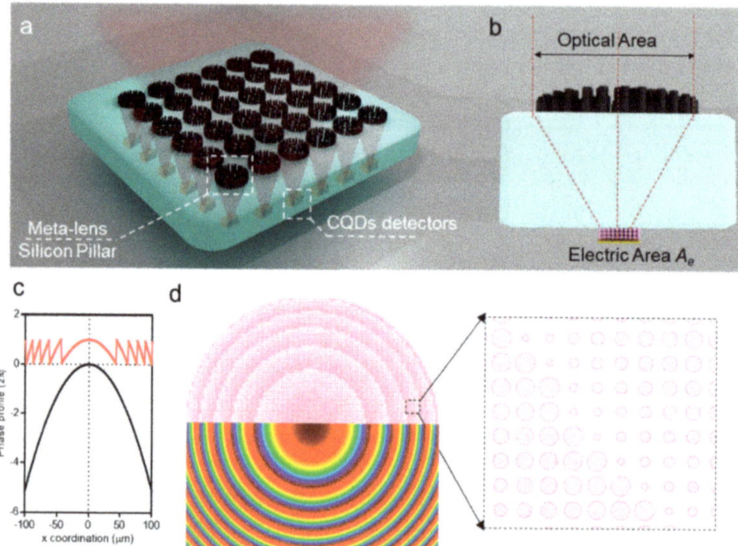

Figure 2. (**a**) Illustration of meta-lens integrated with CQD detectors. Incident light is collected and focused into CQD detectors with a small electric area. (**b**) Cross-sectional illustration of the meta-lens integrated with the CQD detector. Light illuminated onto the optical area is focused to the CQD detector with electric area A_e. (**c**) Phase profile of a regular lens (black curve) and meta-lens (red). (**d**) Design of meta-lens using pillar structures. The diameter and focal length are 200 and 200 μm, respectively. Inset is a representative 2D phase mapping used to generate the meta-lens design.

To verify the optical function of the meta-lens design, the light intensity distribution after passing through the meta-lens is simulated. Various values of numerical aperture (NA) were used and the results showed that the incident light could be successfully concentrated and focused into a smaller spot and that the focal length could be precisely tuned by changing the geometrical parameters of the meta-lens (Figure 3a). For a meta-lens with diameter of 200 μm, the diameter of the focused region is around 10–20 μm, depending on the NA (Figure 3b). Overall, the enhancement ratio of light intensity in the focused region is ~20, which is similar to previously reported experimental values [26].

Besides simulation of the meta-lens with different NAs, chromatic aberration was also investigated. For the meta-lens with a NA of 0.43 and targeted wavelength of 4 μm, we calculated the intensity distribution for incident light with different wavelengths. Chromatic aberration was clearly found, as shown in Figure 4a. As the wavelength deviates from the targeted center wavelength of 4 μm, the focal spot obviously shifts, moving the focused light away from the designed position where the CQD detector would be located. Chromatic aberration limits the operation spectral ranges of the meta-lens and is hard to correct. However, such properties could be potentially used to develop infrared detectors with spectral selectivity (Figure 4b). Spectral selectivity presents in all three meta-lenses (2.5, 4.0, and 4.8 μm). Away from the targeted wavelength, the enhancement ratio decreases rapidly. The simulation results clearly demonstrate the versatility and flexibility of meta-lenses.

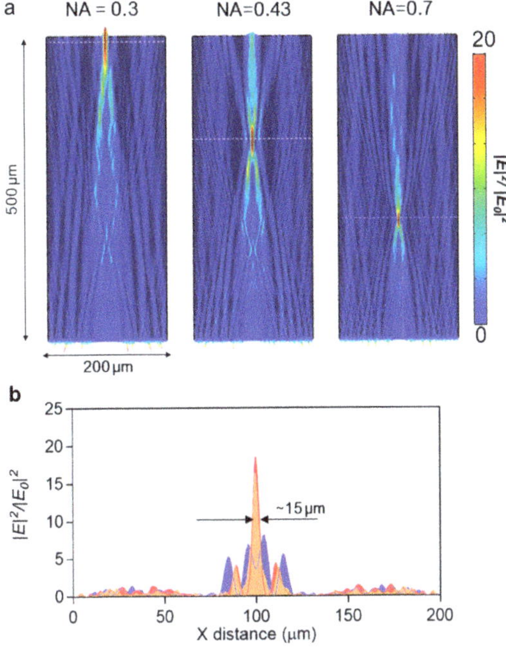

Figure 3. (**a**) COMSOL simulation of light intensity distribution with various numerical apertures (NAs). (**b**) Enhancement ratio profile across the focused area (white dashed line in Figure 2a).

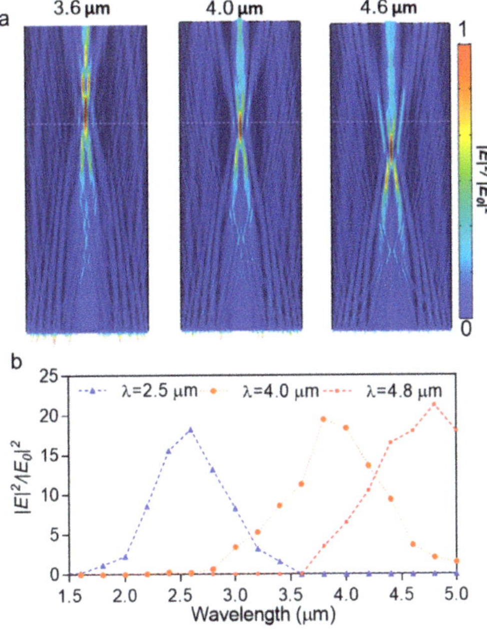

Figure 4. (**a**) Simulated light intensity distribution of a meta-lens with NA of 0.43 with light at different wavelength. (**b**) Spectral enhancement factors across the focused area for meta-lens designed for 2.5, 4.0, and 4.8 μm (dashed line in Figure 3a).

Finally, we simulated the full CQD device stacks with a meta-lens using COMOSL. The simulation model is shown in Figure 5a. The incident light is first focused by the meta-lens and then absorbed by the CQD layers. The diameter of the meta-lens is 200 μm and the width of the CQD detector is swept from 5 to 30 μm with the same thickness of 400 nm. As shown in the simulated light intensity distribution mapping, most of the incident light can be focused onto the CQD detectors by tuning the geometrical parameters of the meta-lens (Figure 5b). With the width of the CQD detectors increasing from 5 to 30 μm, the absorption gradually increases from 10% to 80% and saturates at a width of 25 μm (Figure 5c). Compared to the reference detector with a width of 200 μm without a meta-lens, it is impressive that 4 times more light can be absorbed by a smaller detector at the targeted wavelength of ~4 μm. Based on Equation (3) and previous discussion, higher absorption leads to higher responsivity $\mathfrak{R}$ (×4) and smaller detector results in smaller noise current (×0.125). Therefore, the employment of meta-lenses as optical concentrators could improve the detectivity by a factor of 32. Previous experimental studies of HgCdTe with meta-lenses reported an enhancement ratio of ~5.5, demonstrating that the concept of integration of meta-lenses with infrared detectors should work [26]. The current study mainly focuses on simulation and theoretical calculation, and experimental tests would be conducted to verify such an enhancement factor in the future.

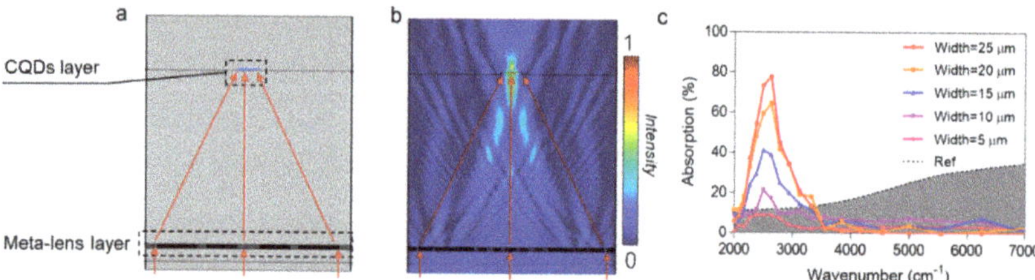

Figure 5. (**a**) Simulation model used to calculate the absorption of CQD detector with meta-lens. (**b**) Simulated light intensity distribution of light passing through meta-lens. (**c**) Simulated absorption of CQD detector with width of 5 to 25 μm with meta-lens and simulated absorption from a reference detector with a width of 200 μm (no meta-lens).

4. Conclusions

In conclusion, our simulation results show that integration of meta-lenses with CQD detectors is a promising way to improve detector detectivity by increasing absorption and decreasing noise at the same time. Full device stack simulation shows that, with meta-lenses, a high absorption of 80% can be achieved even when the electric area of the CQD detectors is decreased by a factor of 64. With meta-lenses as optical concentrators, the detectivity can be improved by a factor of 32. By drop-casting HgTe CQDs on a commercial read-out integrated circuit, a previous study demonstrated that a MWIR thermal camera with a noise equivalent differential temperature (NEDT) of 102 mK [27] showed a simple fabrication route to low-cost thermal cameras with CQDs. Therefore, we believe that integration of CQD infrared detectors with meta-lenses could serve as a promising route towards both high performance single-element detectors and focal plane arrays.

Author Contributions: Conceptualization, X.T.; Methodology, Y.N., S.Z., and Y.H.; Supervision, X.T. and Q.H.; All authors have read and agreed to the published version of the manuscript.

Funding: This work was supported by the National Science Foundation of China (NSFC) (62035004).

Conflicts of Interest: The authors declare no conflict of interest.

References

1. Rogalski, A. History of infrared detectors. *Opto-Electron. Rev.* **2012**, *20*, 279–308. [CrossRef]
2. Rogalski, A. Toward third generation HgCdTe infrared detectors. *J. Alloys Compd.* **2004**, *371*, 53–57. [CrossRef]

3. Böberl, M.; Kovalenko, M.V.; Gamerith, S.; List, E.J.W.; Heiss, W. Inkjet-printed nanocrystal photodetectors operating up to 3 μm wavelengths. *Adv. Mater.* **2007**, *19*, 3574–3578. [CrossRef]
4. Jagtap, A.; Goubet, N.; Livache, C.; Chu, A.; Martinez, B.; Gréboval, C.; Qu, J.; Dandeu, E.; Becerra, L.; Witkowski, N.; et al. Short wave infrared devices based on HgTe nanocrystals with air stable performances. *J. Phys. Chem. C* **2018**, *122*, 14979–14985. [CrossRef]
5. Ackerman, M.M.; Tang, X.; Guyot-Sionnest, P. Fast and sensitive colloidal quantum dot mid-wave infrared photodetectors. *ACS Nano* **2018**, *12*, 7264–7271. [CrossRef] [PubMed]
6. Tang, X.; Ackerman, M.M.; Guyot-Sionnest, P. Thermal imaging with plasmon resonance enhanced HgTe colloidal quantum dot photovoltaic devices. *ACS Nano* **2018**, *12*, 7362–7370. [CrossRef] [PubMed]
7. Keuleyan, S.; Lhuillier, E.; Guyot-Sionnest, P. Synthesis of colloidal HgTe quantum dots for narrow mid-IR emission and detection. *J. Am. Chem. Soc.* **2011**, *133*, 16422–16424. [CrossRef] [PubMed]
8. Keuleyan, S.; Lhuillier, E.; Brajuskovic, V.; Guyot-Sionnest, P. Mid-infrared HgTe colloidal quantum dot photodetectors. *Nat. Photonics* **2011**, *5*, 489–493. [CrossRef]
9. Guyot-Sionnest, P.; Roberts, J.A. Background limited mid-infrared photodetection with photovoltaic HgTe colloidal quantum dots. *Appl. Phys. Lett.* **2015**, *107*, 253104. [CrossRef]
10. Keuleyan, S.E.; Guyot-Sionnest, P.; Delerue, C.; Allan, G. Mercury telluride colloidal quantum dots: Electronic structure, size-dependent spectra, and photocurrent detection up to 12 μm. *ACS Nano* **2014**, *8*, 8676–8682. [CrossRef]
11. Tang, X.; Wu, G.F.; Lai, K.W.C. Plasmon resonance enhanced colloidal HgSe quantum dot filterless narrowband photodetectors for mid-wave infrared. *J. Mater. Chem. C* **2017**, *5*, 362–369. [CrossRef]
12. Goubet, N.; Jagtap, A.; Livache, C.; Martinez, B.; Portalès, H.; Xu, X.Z.; Lobo, R.P.S.M.; Dubertret, B.; Lhuillier, E. Terahertz HgTe nanocrystals: Beyond confinement. *J. Am. Chem. Soc.* **2018**, *140*, 5033–5036. [CrossRef] [PubMed]
13. Chen, M.; Yu, H.; Kershaw, S.V.; Xu, H.; Gupta, S.; Hetsch, F.; Rogach, A.L.; Zhao, N. Fast, Air-stable infrared photodetectors based on spray-deposited aqueous HgTe quantum dots. *Adv. Funct. Mater.* **2014**, *24*, 53–59. [CrossRef]
14. Lhuillier, E.; Robin, A.; Ithurria, S.; Aubin, H.; Dubertret, B. Electrolyte-gated colloidal nanoplatelets-based phototransistor and its use for bicolor detection. *Nano Lett.* **2014**, *14*, 2715–2719. [CrossRef] [PubMed]
15. Huo, N.; Gupta, S.; Konstantatos, G. MoS_2–HgTe quantum dot hybrid photodetectors beyond 2 μm. *Adv. Mater.* **2017**, *29*, 1606576. [CrossRef]
16. Tang, X.; Ackerman, M.M.; Shen, G.; Guyot-Sionnest, P. Towards infrared electronic eyes: Flexible colloidal quantum dot photovoltaic detectors enhanced by resonant cavity. *Small* **2019**, *15*, 1804920. [CrossRef]
17. Tang, X.; Ackerman, M.M.; Guyot-Sionnest, P. Acquisition of hyperspectral data with colloidal quantum dots. *Laser Photon. Rev.* **2019**, *13*, 1900165. [CrossRef]
18. Tang, X.; Chen, M.; Kamath, A.; Ackerman, M.M.; Guyot-Sionnest, P. Colloidal quantum-dots/graphene/silicon dual-channel detection of visible light and short-wave infrared. *ACS Photonics* **2020**, *7*, 1117–1121. [CrossRef]
19. Tang, X.; Ackerman, M.M.; Chen, M.; Guyot-Sionnest, P. Dual-band infrared imaging using stacked colloidal quantum dot photodiodes. *Nat. Photonics* **2019**, *13*, 277–282. [CrossRef]
20. Tang, X.; Chen, M.; Ackerman, M.M.; Melnychuk, C.; Guyot-Sionnest, P. Direct imprinting of quasi-3D nanophotonic structures into colloidal quantum-dot devices. *Adv. Mater.* **2020**, *32*, 1906590. [CrossRef]
21. Gedvilas, M.; Voisiat, B.; Indrišiūnas, S.; Račiukaitis, G.; Veiko, V.; Zakoldaev, R.; Sinev, D.; Shakhno, E. Thermo-chemical microstructuring of thin metal films using multi-beam interference by short (nano- & picosecond) laser pulses. *Thin Solid Films* **2017**, *634*, 134–140. [CrossRef]
22. Zhang, Z.; Dong, L.; Ding, Y.; Li, L.; Weng, Z.; Wang, Z. Micro and nano dual-scale structures fabricated by amplitude modulation in multi-beam laser interference lithography. *Opt. Express* **2017**, *25*, 29135. [CrossRef]
23. Zou, X.; Zheng, G.; Yuan, Q.; Zang, W.; Chen, R.; Li, T.; Li, L.; Wang, S.; Wang, Z.; Zhu, S. Imaging based on metalenses. *PhotoniX* **2020**, *1*, 1–24. [CrossRef]
24. Huang, T.Y.; Grote, R.R.; Mann, S.A.; Hopper, D.A.; Exarhos, A.L.; Lopez, G.G.; Kaighn, G.R.; Garnett, E.C.; Bassett, L.C. A monolithic immersion metalens for imaging solid-state quantum emitters. *Nat. Commun.* **2019**, *10*, 1–8. [CrossRef] [PubMed]
25. Dharmavarapu, R.; Hock Ng, S.; Eftekhari, F.; Juodkazis, S.; Bhattacharya, S. MetaOptics: Opensource software for designing metasurface optical element GDSII layouts. *Opt. Express* **2020**, *28*, 3505. [CrossRef] [PubMed]

26. Li, F.; Deng, J.; Zhou, J.; Chu, Z.; Yu, Y.; Dai, X.; Guo, H.; Chen, L.; Guo, S.; Lan, M.; et al. HgCdTe mid-Infrared photo response enhanced by monolithically integrated meta-lenses. *Sci. Rep.* **2020**, *10*, 1–10. [CrossRef] [PubMed]
27. Buurma, C.; Pimpinella, R.E.; Ciani, A.J.; Feldman, J.S.; Grein, C.H.; Guyot-Sionnest, P. MWIR imaging with low cost colloidal quantum dot films. In Proceedings of the Optical Sensing, Imaging, and Photon Counting: Nanostructured Devices and Applications, San Diego, CA, USA, 8 December 2016; Razeghi, M., Temple, D.S., Brown, G.J., Eds.; International Society for Optics and Photonics: Bellingham, WA, USA, 2016; Volume 9933, p. 993303.

Article

Simulation and Design of HgSe Colloidal Quantum-Dot Microspectrometers

Chong Wen [1], Xue Zhao [1], Ge Mu [1], Menglu Chen [1,2,3] and Xin Tang [1,2,3,*]

1 School of Optics and Photonics, Beijing Institute of Technology, Beijing 100081, China; 3120200645@bit.edu.cn (C.W.); 3220210544@bit.edu.cn (X.Z.); 7520210145@bit.edu.cn (G.M.); menglu@bit.edu.cn (M.C.)
2 Beijing Key Laboratory for Precision Optoelectronic Measurement Instrument and Technology, Beijing 100081, China
3 Yangtze Delta Region Academy of Beijing Institute of Technology, Jiaxing 314019, China
* Correspondence: xintang@bit.edu.cn

Abstract: In recent years, colloidal quantum dots (CQD) have been intensively studied in various fields due to their excellent optical properties, such as size-tunable absorption features and wide spectral tunability. Therefore, CQDs are promising infrared materials to become alternatives for epitaxial semiconductors, such as HgCdTe, InSb, and type II superlattices. Here, we report a simulation study of a microspectrometer fabricated by integrating an intraband HgSe CQD detector with a distributed Bragg reflector (DBR). Intraband HgSe CQDs possess unique narrowband absorption and optical response, which makes them an ideal material platform to achieve high-resolution detection for infrared signatures, such as molecular vibration. A microspectrometer with a center wavelength of 4 μm is studied. The simulation results show that the optical absorption rate of the HgSe CQD detector can be increased by 300%, and the full-width-at-half-maximum (FWHM) is narrowed to 30%, realizing precise regulation of the absorption wavelength. The influence of the incident angle of light waves on the microspectrometer is also simulated, and the results show that the absorption rate of the HgSe quantum dot detector is increased 2–3 times within the incident angle of 0–23 degrees, reaching a spectral absorption rate of more than 80%. Therefore, we believe that HgSe CQDs are a promising material for realizing practical HgSe microspectrometers.

Keywords: colloidal quantum dots; distributed bragg reflector; microspectrometer

Citation: Wen, C.; Zhao, X.; Mu, G.; Chen, M.; Tang, X. Simulation and Design of HgSe Colloidal Quantum-Dot Microspectrometers. *Coatings* **2022**, *12*, 888. https://doi.org/10.3390/coatings12070888

Academic Editor: Aomar Hadjadj

Received: 8 April 2022
Accepted: 13 June 2022
Published: 22 June 2022

1. Introduction

Spectral detection is the core technology used in the fields of physicochemical property analysis, chemical composition analysis, and environmental monitoring [1–4]. It also shows great potential in emerging artificial intelligence networks, autonomous driving, and machine vision [5–7]. A spectrometer is generally composed of dispersive optics, a detector system, and an optical path adjustment system. An important development direction of the spectrometer is the miniaturization of spectrometers. Micro-spectrometers with a small size, low energy consumption, and low cost have become promising alternatives in many spectral detection fields. Currently, micro-spectrometers are mainly made by micro-nano optical processing or micro-electro-mechanical system (MEMS) technology, including photonic crystal arrays [8–10], arrayed waveguide gratings [11], planar etched gratings [12], micro-Michelson interferometers [13], and narrowband optical filters [14,15]. The use of non-silicon-based photodetectors is another way to fabricate spectrometers, including graphene detectors [16], organic photodetectors [17], and colloidal quantum dot detectors (CQDs) [18]. Compared with MEMS and etched gratings, spectrometers fabricated with detectors capable of detecting specific wavelengths have the advantages of compact structure and system stability.

Infrared detectors play a key role in spectrometers. Colloidal quantum dots (CQD) have been intensively studied in various fields due to their excellent optical properties, such as size-tunable absorption features and wide spectral tunability. Therefore, CQDs are promising infrared materials to become alternatives for epitaxial semiconductors, such as HgCdTe, InSb, and type II superlattices. At present, the tunable bands of the developed CQDs have covered the short-wave infrared (SWIR) [19], mid-wave infrared (MWIR) [20], long-wave infrared (LWIR), and terahertz bands [21,22]. Photodetectors based on quantum dots for photoconductivity [23], photovoltaics [20,24], and phototransistors have been investigated [25]. Benefiting from the ease of processing and spectral tunability of CQDs, filter-free narrowband photodetectors [26], double-endedSWIR/MWIR dual-band detectors [27], and SWIR hyperspectral sensors have been investigated [28].

Among narrow-band optoelectronic materials, mercury selenide (HgSe) CQDs have thus far demonstrated the only Lorentzian lineshape of optical absorption in mid-wave infrared (MWIR), which is suitable for spectrally selective detection [29,30]. Despite the exciting properties of HgSe CQDs, most infrared CQD detectors still suffer from a common problem, a low absorption rate limited by the CQD film thickness [24]. CQD detectors are usually made by layer-by-layer drop-casting, and for an infrared CQD detector, the typical thickness is ~400 nm, and only 20% of incident light can be absorbed [31]. Another problem that needs to be improved is that HgSe CQDs still have a relatively broad FWHM. In this work, high-performance HgSe CQD microspectrometers are designed by integrating a HgSe CQD sensor and a resonant cavity (RC) composed of distributed Bragg reflectors (DBR). Finite element analysis of HgSe CQD microspectrometer detectors was systematically conducted.

The simulation results show that the spectrometer has strong wavelength selectivity, high absorption rate up to 91%, and extremely narrow full width at half maximum down to 0.2 μm in MWIR. The wavelength selection of the absorption peak can be achieved by adjusting the thickness of the optical cavity. Therefore, the integration of the narrowband CQD material HgSe with the resonator can achieve precise spectrally selective detection. It can be concluded that the spectrometer has the advantages of adjustable wavelength, narrow half-peak, high spectral resolution, high spectral responsivity, small size, and low cost.

2. Modeling and Simulation

COMSOL Multiphysics is used to conduct the simulation in our study. The proposed design of a CQD microspectrometer with a resonant cavity is illustrated in Figure 1a. A sapphire (Al_2O_3) with high transmittance in the infrared band is used as the substrate, and Ti_3O_5 and SiO_2 are periodically deposited on it as the DBR. The optical spacer layer and CQD detector were fabricated on the DBR and capped with insulating layers and gold film. The DBR is composed of alternating layers of two materials, each of which has a different refractive index. The interference effect between reflected waves makes the reflection window of the DBR depend on wavelength. To save computation resources, a two-dimensional (2D) simulation is used, as shown in Figure 1b. There is only one physical field in the model, a simple parallel light field entering the device from the entrance port. The light is introduced from the "incident port" with a controlled power of 1 W. The wavelength of incident light is swept from 2.5 to 10.0 μm to calculate spectral absorption.

The listening port is placed at the bottom of the model. Considering that the size of the model is much smaller than the size of the actual device, we set the left and right sides as "continuous periodic boundary conditions" to avoid the influence caused by the limited width of the model. Below the incident port is the substrate, and the material of the substrate is chosen to be Al_2O_3 with a refractive index of 1.6 + 0i because of its low absorption for infrared light.

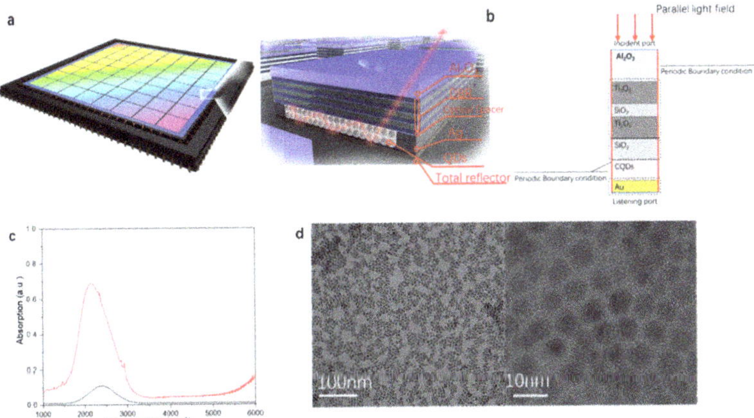

Figure 1. Architecture of the HgSe CQD microspectrometer. (**a**) Schematic of the CQD microspectrometer, consisting of a CQD sensor and filters. The optical spacer layer is under the DBR, and the quantum dot layer is sandwiched between the optical spacer layer and the optical reflection layer. (**b**) Simulation model of a CQD microspectrometer with a resonant cavity. (**c**) Spectral absorption plots of HgSe CQDs (red line), simulated spectral absorption plots (black lines). (**d**) TEM of HgSe CQDs.

The DBR is composed of alternating layers of high and low refractive index dielectric films. The material of the high refractive index layer of the DBR is selected as Ti_3O_5 with a refractive index of 2.35 + 0i, and the material of the low refractive index layer is selected as SiO_2 with a refractive index of 1.4 + 0i. The optical spacer layer is used to regulate the thickness of the resonator cavity, and the material is also SiO_2. HgSe CQD were used as the photosensitive layer of the micro-spectrometer, and the refractive index was obtained by fitting the actual absorption spectrum of HgSe CQD, which had an absorption peak with a central wavelength of 4.19 μm in the range of 1.67–10 μm, as shown in Figure 1c,d. The mean size of the CQDs is 4.7 nm, and the size distribution is uniform, with a standard deviation of 0.5 nm. The bottom of the model is the gold film, as the total reflection layer of light waves, which, together with the DBR and the optical spacer layer, constitutes an optical resonant cavity. The rest of the model is set to air with a refractive index of 1 + 0i. In the simulation, the width of the 2D model is 0.4 μm, the thickness of the Al_2O_3 substrate is 1.1 μm, the thickness of the HgSe CQD film is set to 0.4 μm which is the thickest thickness of the active layer in most CQD detectors [31], and the thickness of the DBR dielectric film layer is determined by [32]:

$$\mathrm{n_H d_H} = \mathrm{n_L d_L} = \lambda_0/4 \tag{1}$$

where $\mathrm{n_H}$ and $\mathrm{n_L}$ are the refractive indices of the DBR high and low refractive index dielectric layers, $\mathrm{d_H}$ and $\mathrm{d_L}$ are the thicknesses of the two dielectric layers, and λ_0 is the center wavelength of the DBR reflection window. In the design, we use unpolarized plane waves with uniform intensity distribution. To reduce manufacturing complexity, the spectrometer uses a vertically stacked architecture. The absorption, wavelength of the absorption peak, and FWHM can be regulated by adjusting the thickness of the optical spacer layer.

3. Results and Discussion

When light passes through the structure, reflected waves are generated between multiple interfaces. The bandwidth of the reflection window of the DBR is given by [33,34]:

$$\Delta\lambda_0 = \frac{4\lambda_0}{\pi}\arcsin\left(\frac{\mathrm{n_H} - \mathrm{n_L}}{\mathrm{n_H} + \mathrm{n_L}}\right) \tag{2}$$

where λ_0 is the center wavelength of the bandwidth. Another important factor determined by the DBR is the reflectivity of the resonator top mirror. The reflectivity R of the distributed DBR is given by:

$$R = \left(\frac{1 - \left(\frac{n_H}{n_L}\right)^{2N} \frac{n_H^2}{n_n n_b}}{1 + \left(\frac{n_H}{n_L}\right)^{2N} \frac{n_H^2}{n_a n_b}} \right)^2 \quad (3)$$

where n_H and n_L are the refractive indices of the high refractive index layer and the low refractive index layer, n_a and n_b are the refractive index of substrates on both sides of the DBR, and N is the number of layers of the DBR. For the resonant cavity, continuous resonance wavelength modulation is achieved by changing the optical length of the layer of the DBR and cavity, which can be done by tuning the thickness of the optical spacer layer and the layer of the DBR. In our design, the thickness of the DBR dielectric layer needs to match the absorption wavelength of the quantum dot narrowband material, so the absorption rate and FWHM of the photodetector source layer can be regulated by simply changing the number of DBR layers and the thickness of the optical isolation layer.

By introducing spectrally flat light as an illumination source, the optical field distributions in the resonant cavity are simulated, as shown in Figure 2. The core of the optical field is located on the boundary between the active layer and the optical spacer layer, which may be due to the absorption of photons by the active layer. Since the active layer placed in the resonant cavity is a narrowband CQD material, unwanted resonances and broadening FWHM of peaks are avoided, allowing tunable narrowband photodetection. The simulated absorption of the microspectrometer further demonstrates the accurate spectral tunability of our device.

The quantum efficiency depicting the absorption of a photodetector to optical signals is also analyzed. At zero bias, assuming that the optical active layer is defect free, η can be calculated using:

$$\eta = (1 - R)\left(1 - e^{-\alpha d}\right) \quad (4)$$

where $\alpha(h\nu)$ is the absorption coefficient, d is the thickness of the active layer, and R is surface reflection of the photodetectors. In our design, the total quantum efficiency η_{tot} of photodetector with RC is [35,36]:

$$\eta_{tot} = \frac{1 + R_2 e^{-\alpha d}}{1 - 2\sqrt{R_1 R_2} e^{-\alpha d} \cos\left(\frac{4n\pi}{\lambda} L + \psi_1 + \psi_2\right) + R_1 R_2 e^{-2\alpha d}} \eta \quad (5)$$

where R_1 is the corresponding power reflectivity of the top mirror composed of DBR of RC, R_2 is the corresponding power reflectivity of the bottom reflective mirror composed of gold reflective film of RC, L, and n are the thickness and refractive index of the optical spacer layer, ψ_1, ψ_2 are the phase shift caused by the top and the bottom mirror of the resonant cavity to the reflected light, and λ is the wavelength of incident light. The improvement in the reflectivity of the top mirror increases the resonant effect of the resonator and hinders light from entering the resonator, so the quantum efficiency of the photodetector is greatly reduced when the number of DBR layers increases, which is proved in our simulation. In truth, in our simulations, when the number of layers of the DBR reaches three, the absorption rate of the CQD detector drops sharply. The full quantum efficiency is more dependent on the αd of the optoelectronic material. Limited by the layer-by-layer drop-casting method of the CQD detector, a further increase of film thickness would lead to cracks and delamination due to mechanical stress from the drying and ligand exchange process [31]. Additionally, the variation in the thickness of the optical isolation layer affects the periodic shift of the central wavelength of the spectral absorption peak.

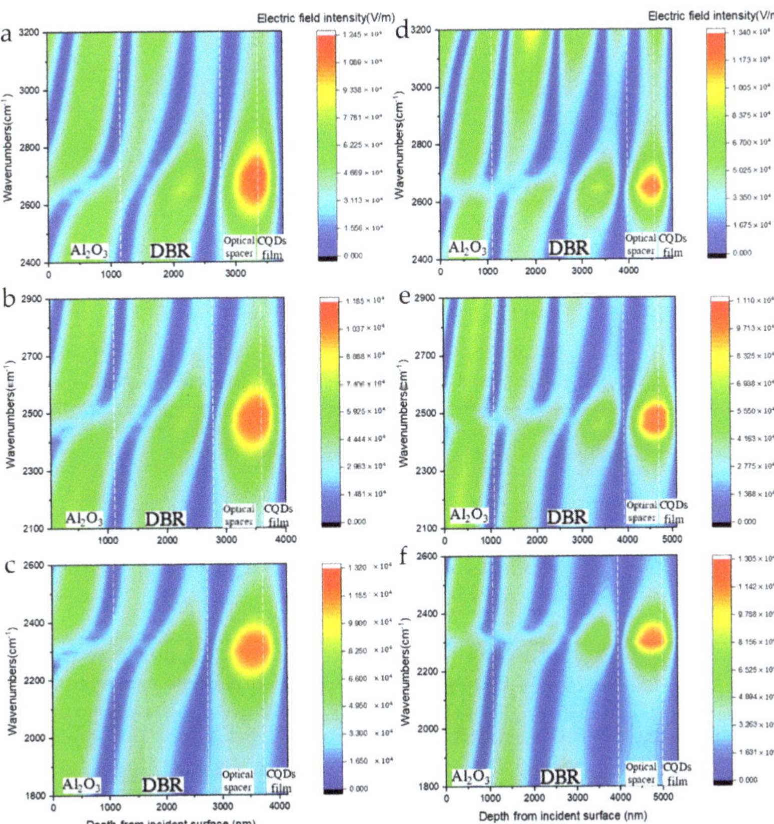

Figure 2. Optical field distributions of the resonant cavity. Optical field distributions for optical spacer layer thicknesses of (**a**) 0.6, (**b**) 0.8, and (**c**) 1.0 μm of one layer of DBR are shown. (**d–f**). The optical field distributions for two layers of DBR with the same optical spacer layer thickness as one layer. Different layer positions are marked by white dashed lines.

The HgSe active layer in the device is replaced by a virtual broadband material whose imaginary part of the refractive index is constant, independent of the wavelength of light, unlike HgSe CQDs. It can be considered a broad-spectrum optoelectronic material with the same absorption for different wavelengths of light, and the results are shown in Figure 3. For broad-spectrum optoelectronic materials, the harmonic density and order of the spectral absorption peaks gradually increase, and the side mode suppression ratio also decreases gradually when the wavelength decreases and the thickness of the optical spacer layer increases. In our design, the absorption suppression of the harmonic sub-peaks of the resonant cavity comes from the unique intraband transition characteristics of HgSe CQDs as the active layer material. A unique intraband transition property of HgSe CQDs is that electronic transitions occur between the first two conduction band states, 1S e and 1P e of HgSe CQDs. When the incident photon energy E_{photon} in the infrared matches the intraband energy gap E_{intra}, the light absorption reaches a maximum value, and if the energy of the incident photon is higher or lower than E_{intra}, the light absorption drops sharply, which creates the Lorentzian lineshape of the absorption properties of HgSe CQDs.

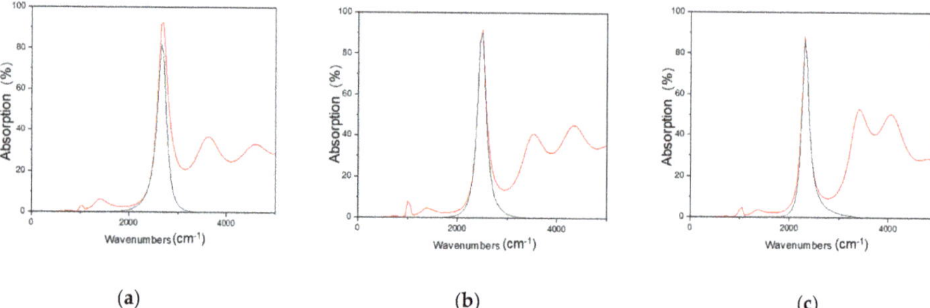

Figure 3. Spectral absorption of photodetectors with resonant cavities constructed from narrowband (HgSe, black line) and broadband (red line) optoelectronic materials at optical spacer layer thicknesses of 0.6 (**a**), 0.8 (**b**), and 1.0 (**c**) μm.

According to existing experimental conclusions, the influence of the size and size of the CQDs on the energy band can be further learned. The absorption peaks of HgSe redshift with increasing size, which is controlled by the reaction conditions for synthesizing CQDs [37,38]. By controlling the reaction time and temperature, the energy gap of the CQDs becomes smaller, and as a result, the Fermi level is in the Se state, defined as the two electrons in the conduction band of each CQD being in the Se state and the Pe state being empty. Compared to other electron doping densities, the dark current of HgSe film is reduced [39,40]. On the other hand, the uneven distribution of quantum dot size will also have an impact on carrier mobility. The carrier mobility of the HgSe film can be measured and calculated by the FET circuit, and the mechanism is explained by Marcus theory [40]:

$$\mu_{EFT} = \mathrm{Con}\exp\left(-(\gamma + \Delta G)^2/4\gamma k_B T\right) \tag{6}$$

where Con is the constant, ΔG is the energy difference caused by uneven size of CQDs, T is the temperature in Kelvin, k_B is the Boltzmann constant. γ is the recombinant energy of the polarization effect of the medium around the carrier [41]. γ is obtained by:

$$\gamma = \frac{e}{4\pi\varepsilon_0}\left(\frac{1}{a} - \frac{1}{2(D+l)}\right)\left(\frac{1}{\varepsilon_1} - \frac{1}{\varepsilon_{st}}\right) \tag{7}$$

where e is the elementary charge, ε_0 is the vacuum permittivity, D is the radius of the CQDs, l is the spacing between CQDs, ε_1 is the optical dielectric constant of the matrix surrounding CQDs, ε_{st} is the static dielectric constant. The intraband transition property makes HgSe CQDs an excellent optoelectronic material for detecting specific wavelengths of infrared. By tuning the size of the HgSe CQDs, the E_{intra} and the corresponding peak absorption wavelength can be controlled. This design can modify the reflection window of the DBR to match the specific wavelength detectable by HgSe by adjusting the optical length of the DBR dielectric thin-film layer. The resonance enhancement effect of the resonant cavity is used to improve the absorption rate and fine-tune the absorption wavelength of the HgSe CQD detector. Previous studies have already demonstrated that photo-excited carriers through intraband transition can be collected to generate photocurrents. The application of the intraband transition for photoconductors was first demonstrated in the 3–5 μm range of infrared [42].

To verify the optical functionality of the DBR design, the spectral absorption of the CQD detector in the single-band microspectrometer was simulated using a one-layer and two-layer DBR, as shown in Figure 4. The number of layers of DBR is determined by the number of high and low refractive index film layer pairs. When the resonant wavelength of the resonant cavity accurately matches the absorption peak of the CQD active layer, the absorption rate of the CQD detector increases 2.8 times. The FWHM is reduced to 18%

and 31%, respectively, in the application of one layer and two layers of DBR. There is an approximate linear relationship between the central wavelength of the absorption peak and the thickness of the optical spacer layer, indicating that the central wavelength of the absorption peak can be precisely tuned by changing the thickness of the optical spacer layer (Figure 4d). Compared with one-layer DBR, the FWHM of the absorption peak of the CQD detector with two layers of DBR is narrower, but the peak absorption oscillates when the optical isolation layer is thickened, which may indicate that increasing the number of layers of DBR will produce more interference when the optical length of the resonant cavity increases.

Overall, the performance of the CQD detector was significantly increased after adopting the resonator with one layer of DBR. The absorption of the CQD detector was increased 2–3 times, and the FWHM was narrowed to 30% of the original. The central wavelength of the absorption peak can be tuned in the range of 0.7 μm, which is consistent with previously reported studies [26].

In addition to modeling the cavity length of the resonator, the influence of the angle of incidence was also studied. For a cavity with an optical spacer layer of 0.82 μm (absorption > 90%, Figure 5a), we simulated spectral absorption of the CQD detector with incident light from 0 degrees to 90 degrees, and the modeling of the incident angle of the light wave is shown in Figure 6a. The simulation results are shown in Figure 6b,c. The results show that within the incident angle of 0–25 degrees, the absorption of the CQD detector is still increased 2–3 times, and the FWHM is narrowed by 30%, but the center wavelength is shifted by 0.3 μm at most (Figure 6d–f). After the incident angle exceeds 25 degrees, the absorption rate of the CQD detector drops sharply, and the center wavelength of the peak of absorption produces a large periodic change.

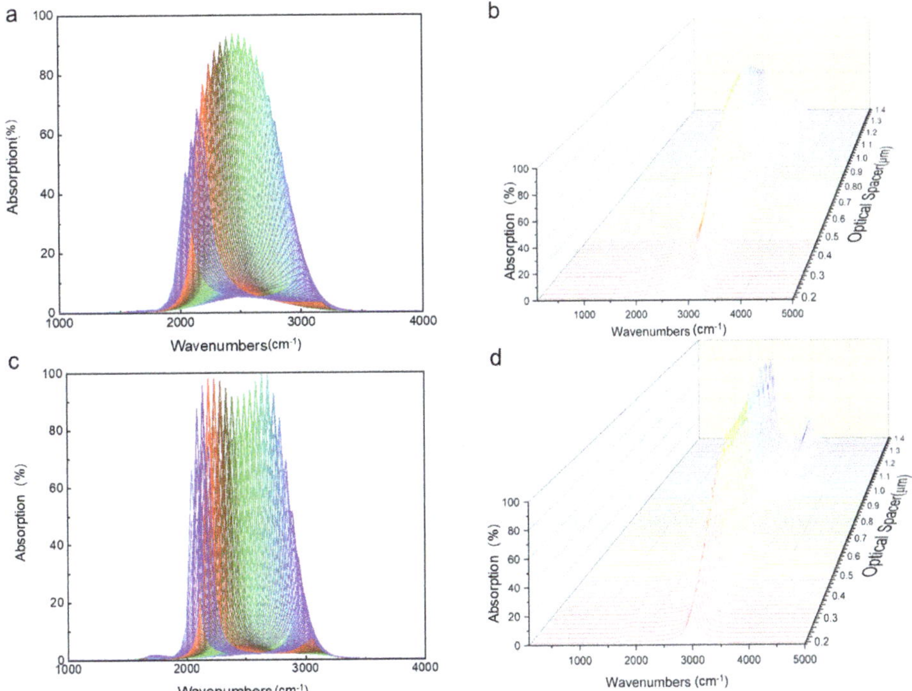

Figure 4. Absorption of the CQD microspectrometer. (**a**,**b**) Absorption of a CQD microspectrometer with different optical spacer thicknesses with one layer of DBR. (**c**,**d**) Absorption of a CQD microspectrometer with different optical spacer thicknesses and two layers of DBR.

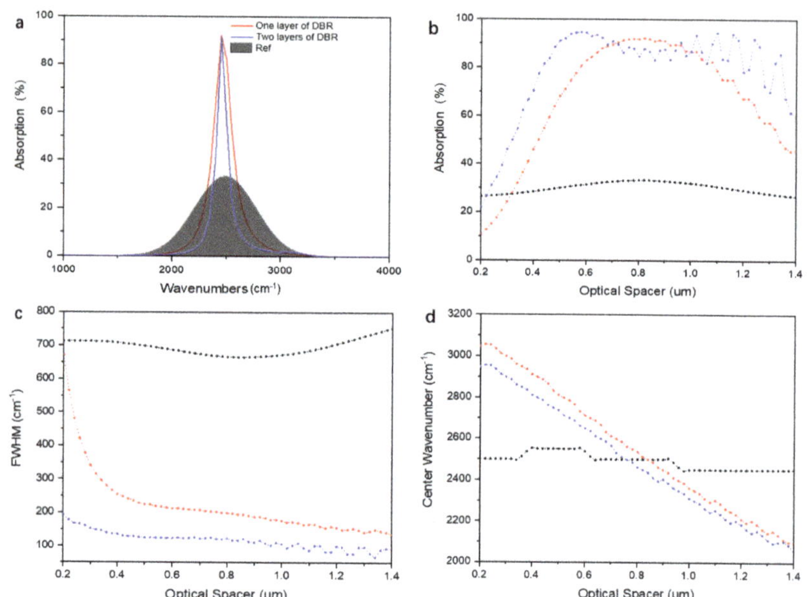

Figure 5. Effect of resonant cavity composed of DBR on CQD detector. (**a**) Absorption of the CQD detector with one layer of DBR, two layers of DBR, and no DBR (Ref) for the thickness of the optical spacer layer is 0.82 μm. Absorption (**b**), FWHW (**c**), and center wavenumber (**d**) of absorption peak of the CQD detector at different thicknesses of the optical spacer layer with one layer (red line), two layers (blue line) of DBR, and no DBR (black line).

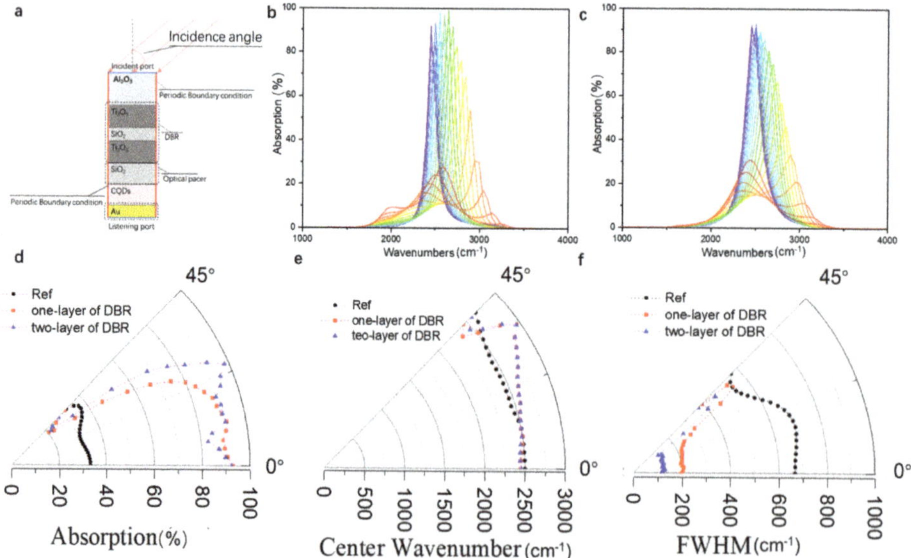

Figure 6. Influence of incidence angle on mincrospectrometer performance. (**a**) Simulation model of a CQD microspectrometer with an incident angle. Absorption of a CQD microspectrometer with different incident angles in one layer of DBR (**b**) and two layers of DBR (**c**). Absorption (**d**), center wavenumber (**e**), and FWHW (**f**) of absorption peak of the CQD detector in the range of incident angle of 0–45 degrees with one layer (red line), two layers (blue line) of DBR, and no DBR (black line).

4. Conclusions

In conclusion, our design and simulation confirm that the wavelength-specific detection of HgSe for infrared forms a good match with the reflection window formed by DBR and the resonance wavelength of the resonant cavity, which can further obtain a lower FWHM and a higher absorption rate. The detector assumes the function of a single-band spectrometer in the range of MWIR according to the simulation results, and it is possible to fabricate a snapshot HgSe CQD microspectrometer by integrating multiple photodetectors capable of detecting various infrared regions by simply adjusting the cavity length of the resonant cavity. We believe that the simulation results have important implications for microspectrometers. In addition, the effect of enhancing the light absorption of the active layer with this device structure should be applicable to thinner film-like detectors.

Author Contributions: Conceptualization, X.T., C.W. and G.M.; methodology, M.C.; software, C.W.; validation, X.T., C.W., G.M. and X.Z.; formal analysis, X.Z. and M.C.; investigation, X.T., C.W. and G.M.; resources, X.T., C.W. and G.M.; data curation, X.T., C.W., X.Z. and G.M.; writing—original draft preparation, C.W. and G.M.; writing—review and editing, X.T. and M.C.; visualization, X.T., C.W. and G.M.; supervision, X.T.; project administration, X.T.; funding acquisition, X.T. All authors have read and agreed to the published version of the manuscript.

Funding: This research was funded by the National Key R&D Program of China: 2021YFA0717600; National Natural Science Foundation of China: NSFC No. 62035004; National Natural Science Foundation of China: NSFC No. 62105022.

Institutional Review Board Statement: Not applicable.

Informed Consent Statement: Not applicable.

Data Availability Statement: Not applicable.

Conflicts of Interest: The authors declare no conflict of interest.

References

1. Kindt, J.T.; Luchansky, M.S.; Qavi, A.J.; Lee, S.-H.; Bailey, R.C. Subpicogram Per Milliliter Detection of Interleukins Using Silicon Photonic Microring Resonators and an Enzymatic Signal Enhancement Strategy. *Anal. Chem.* **2013**, *85*, 10653–10657. [CrossRef] [PubMed]
2. Ryckeboer, E.; Bockstaele, R.; Vanslembrouck, M.; Baets, R. Glucose sensing by waveguide-based absorption spectroscopy on a silicon chip. *Biomed. Opt. Express* **2014**, *5*, 1636–1648. [CrossRef] [PubMed]
3. Garcia-Perez, A.; Romero-Troncoso, R.; Cabal-Yepez, E.; Osornio-Rios, A.R. The Application of High-Resolution Spectral Analysis for Identifying Multiple Combined Faults in Induction Motors. *IEEE Trans. Ind. Electron.* **2011**, *58*, 2002–2010. [CrossRef]
4. Green, O.R.; Eastwood, M.L.; Sarture, C.M.; Chrien, T.G.; Aronsson, M.; Chippendale, B.J.; Faust, A.J.; Pavri, E.B.; Chovit, C.J.; Solis, M.; et al. Imaging Spectroscopy and the Airborne Visible/Infrared Imaging Spectrometer (AVIRIS). *Remote Sens. Environ.* **1998**, *65*, 227–248. [CrossRef]
5. Millán, M.S.; Escofet, J. Fabric inspection by near-infrared machine vision. *Opt. Lett.* **2004**, *29*, 1440–1442. [CrossRef]
6. Janssens, O.; Van De Walle, R.; Loccufier, M.; Van Hoecke, S. Deep Learning for Infrared Thermal Image Based Machine Health Monitoring. *IEEE ASME Trans. Mechatron.* **2017**, *23*, 151–159. [CrossRef]
7. Wang, H.; Wang, B.; Liu, B.; Meng, X.; Yang, G. Pedestrian recognition and tracking using 3D LiDAR for autonomous vehicle. *Robot. Auton. Syst.* **2017**, *88*, 71–78. [CrossRef]
8. Ding, H.; Liu, C.; Gu, H.; Zhao, Y.; Wang, B.; Gu, Z. Responsive Colloidal Crystal for Spectrometer Grating. *ACS Photon.* **2014**, *1*, 121–126. [CrossRef]
9. Kim, S.-H.; Park, H.S.; Choi, J.H.; Shim, J.W.; Yang, S.-M. Integration of Colloidal Photonic Crystals toward Miniaturized Spectrometers. *Adv. Mater.* **2009**, *22*, 946–950. [CrossRef]
10. Redding, B.; Liew, S.F.; Sarma, R.; Cao, H. Compact spectrometer based on a disordered photonic chip. *Nat. Photon.* **2013**, *7*, 746–751. [CrossRef]
11. Muneeb, M.; Vasiliev, A.; Ruocco, A.; Malik, A.; Chen, H.; Nedeljkovic, M.; Penades, J.S.; Cerutti, L.; Rodriguez, J.-B.; Mashanovich, G.Z.; et al. III-V-on-silicon integrated micro—spectrometer for the 3 μm wavelength range. *Opt. Express* **2016**, *24*, 9465–9472. [CrossRef]
12. Kyotoku, B.B.C.; Chen, L.; Lipson, M. Sub-nm resolution cavity enhanced microspectrometer. *Opt. Express* **2009**, *18*, 102–107. [CrossRef] [PubMed]
13. Salem, A.M.; Sabry, Y.M.; Fathy, A.; Khalil, D.A. Single MEMS Chip Enabling Dual Spectral-Range Infrared Micro-Spectrometer with Optimal Detectors. *Adv. Mater. Technol.* **2021**, *6*, 2001013. [CrossRef]

14. Bao, J.; Bawendi, M.G. A colloidal quantum dot spectrometer. *Nature* **2015**, *523*, 67–70. [CrossRef]
15. Zhu, X.; Bian, L.; Fu, H.; Wang, L.; Zou, B.; Dai, Q.; Zhong, H. Broadband perovskite quantum dot spectrometer beyond human visual resolution. *Light Sci. Appl.* **2020**, *9*, 73. [CrossRef]
16. Liu, Y.; Gong, T.; Zheng, Y.; Wang, X.; Xu, J.; Ai, Q.; Guo, J.; Huang, W.; Zhou, S.; Liu, Z.; et al. Ultra-sensitive and plasmon-tunable graphene photodetectors for micro-spectrometry. *Nanoscale* **2018**, *10*, 20013–20019. [CrossRef] [PubMed]
17. Xing, S.; Nikolis, V.C.; Kublitski, J.; Guo, E.; Jia, X.; Wang, Y.; Spoltore, D.; Vandewal, K.; Kleemann, H.; Benduhn, J.; et al. Miniaturized VIS-NIR Spectrometers Based on Narrowband and Tunable Transmission Cavity Organic Photodetectors with Ultrahigh Specific Detectivity above 10 Jones. *Adv. Mater.* **2021**, *33*, 2102967. [CrossRef]
18. Zobenica, Ž.; Van Der Heijden, R.W.; Petruzzella, M.; Pagliano, F.; Leijssen, R.; Xia, T.; Midolo, L.; Cotrufo, M.; Cho, Y.; Van Otten, F.W.M.; et al. Integrated nano-opto-electro-mechanical sensor for spectrometry and nanometrology. *Nat. Commun.* **2017**, *8*, 2216. [CrossRef]
19. Böberl, M.; Kovalenko, M.; Gamerith, S.; List, E.J.W.; Heiss, W. Inkjet-Printed Nanocrystal Photodetectors Operating up to 3 μm Wavelengths. *Adv. Mater.* **2007**, *19*, 3574–3578. [CrossRef]
20. Ackerman, M.M.; Tang, X.; Guyot-Sionnest, P. Fast and Sensitive Colloidal Quantum Dot Mid-Wave Infrared Photodetectors. *ACS Nano* **2018**, *12*, 7264–7271. [CrossRef]
21. Keuleyan, S.E.; Guyot-Sionnest, P.; Delerue, C.; Allan, G. Mercury Telluride Colloidal Quantum Dots: Electronic Structure, Size-Dependent Spectra, and Photocurrent Detection up to 12 μm. *ACS Nano* **2014**, *8*, 8676–8682. [CrossRef] [PubMed]
22. Goubet, N.; Jagtap, A.; Livache, C.; Martinez, B.; Portalès, H.; Xu, X.Z.; Lobo, R.P.S.M.; Dubertret, B.; Lhuillier, E. Terahertz HgTe Nanocrystals: Beyond Confinement. *J. Am. Chem. Soc.* **2018**, *140*, 5033–5036. [CrossRef] [PubMed]
23. Keuleyan, S.; Lhuillier, E.; Guyot-Sionnest, P. Synthesis of Colloidal HgTe Quantum Dots for Narrow Mid-IR Emission and Detection. *J. Am. Chem. Soc.* **2011**, *133*, 16422–16424. [CrossRef] [PubMed]
24. Tang, X.; Ackerman, M.M.; Guyot-Sionnest, P. Thermal Imaging with Plasmon Resonance Enhanced HgTe Colloidal Quantum Dot Photovoltaic Devices. *ACS Nano* **2018**, *12*, 7362–7370. [CrossRef] [PubMed]
25. Huo, N.; Gupta, S.; Konstantatos, G. MoS2-HgTe Quantum Dot Hybrid Photodetectors beyond 2 μm. *Adv. Mater.* **2017**, *29*, 1606576. [CrossRef] [PubMed]
26. Tang, X.; Wu, G.F.; Lai, K.W.C. Plasmon resonance enhanced colloidal HgSe quantum dot filterless narrowband photodetectors for mid-wave infrared. *J. Mater. Chem. C* **2016**, *5*, 362–369. [CrossRef]
27. Tang, X.; Ackerman, M.M.; Chen, M.; Guyot-Sionnest, P. Dual-band infrared imaging using stacked colloidal quantum dot photodiodes. *Nat. Photon.* **2019**, *13*, 277–282. [CrossRef]
28. Tang, X.; Ackerman, M.M.; Guyot-Sionnest, P. Acquisition of Hyperspectral Data with Colloidal Quantum Dots. *Laser Photon. Rev.* **2019**, *13*, 1900165. [CrossRef]
29. Guyot-Sionnest, P.; Hines, M.A. Intraband transitions in semiconductor nanocrystals. *Appl. Phys. Lett.* **1998**, *72*, 686–688. [CrossRef]
30. Wehrenberg, B.L.; Wang, C.; Guyot-Sionnest, P. Interband and Intraband Optical Studies of PbSe Colloidal Quantum Dots. *J. Phys. Chem. B* **2002**, *106*, 10634–10640. [CrossRef]
31. Tang, X.; Chen, M.; Ackerman, M.M.; Melnychuk, C.; Guyot-Sionnest, P. Direct Imprinting of Quasi-3D Nanophotonic Structures into Colloidal Quantum-Dot Devices. *Adv. Mater.* **2020**, *32*, e1906590. [CrossRef] [PubMed]
32. Bing-Zheng, D.; Jing-Ping, Z.; Yu-Zheng, M.; Hong, L.; Kai, W.; Xun, H. Effects of Bragg periods per grating period on performance of Bragg concave diffraction grating. *Acta Phys. Sin.* **2017**, *66*, 224202. [CrossRef]
33. Zhang, Y.; Zhang, S.-C.; Zhang, H.-B.; Xin, Q.; Kong, Y.-Y.; Chai, B. Comparitive Study of Coaxial Bragg Resonators. *J. Infrared Millim. Terahertz Waves* **2010**, *31*, 1126–1135. [CrossRef]
34. Kang, Y.M.; Arbabi, A.; Goddard, L.L. A microring resonator with an integrated Bragg grating: A compact replacement for a sampled grating distributed Bragg reflector. *Opt. Quantum Electron.* **2009**, *41*, 689–697. [CrossRef]
35. Cui, H.; Zhou, S.; Ma, R.; Huang, Y.; Ren, X. Optimized Design for Resonant-cavity-enhanced (RCE) Photodetector. *Semicond. Optoelectron.* **2005**, *26*, 280–283.
36. Kishino, K.; Unlu, M.S.; Chyi, J.I.; Reed, J.; Arsenault, L.; Morkoc, H. Resonant cavity-enhanced (RCE) photodetectors. *IEEE J. Quantum Electron.* **1991**, *27*, 2025–2034. [CrossRef]
37. Grigel, V.; Sagar, L.K.; De Nolf, K.; Zhao, Q.; Vantomme, A.; De Roo, J.; Infante, I.; Hens, Z. The Surface Chemistry of Colloidal HgSe Nanocrystals, toward Stoichiometric Quantum Dots by Design. *Chem. Mater.* **2018**, *30*, 7637–7647. [CrossRef]
38. Martinez, B.; Livache, C.; Meriggio, E.; Xu, X.Z.; Cruguel, H.; Lacaze, E.; Proust, A.; Ithurria, S.; Silly, M.G.; Cabailh, G.; et al. Polyoxometalate as Control Agent for the Doping in HgSe Self-Doped Nanocrystals. *J. Phys. Chem. C* **2018**, *122*, 26680–26685. [CrossRef]
39. Melnychuk, C.; Guyot-Sionnest, P. Auger Suppression in n-Type HgSe Colloidal Quantum Dots. *ACS Nano* **2019**, *13*, 10512–10519. [CrossRef]
40. Zhao, X.; Mu, G.; Tang, X.; Chen, M. Mid-IR Intraband Photodetectors with Colloidal Quantum Dots. *Coatings* **2022**, *12*, 467. [CrossRef]

41. Prodanović, N.; Vukmirović, N.; Ikonić, Z.; Harrison, P.; Indjin, D. Importance of Polaronic Effects for Charge Transport in CdSe Quantum Dot Solids. *J. Phys. Chem. Lett.* **2014**, *5*, 1335–1340. [CrossRef] [PubMed]
42. Deng, Z.; Jeong, K.S.; Guyot-Sionnest, P. Colloidal Quantum Dots Intraband Photodetectors. *ACS Nano* **2014**, *8*, 11707–11714. [CrossRef] [PubMed]

Article

Simulation of Resonant Cavity-Coupled Colloidal Quantum-Dot Detectors with Polarization Sensitivity

Pengfei Zhao [1,†], Ge Mu [1,†], Menglu Chen [1,2,3] and Xin Tang [1,2,3,*]

1 School of Optics and Photonics, Beijing Institute of Technology, Beijing 100081, China; 3120210617@bit.edu.cn (P.Z.); 7520210145@bit.edu.cn (G.M.); menglu@bit.edu.cn (M.C.)
2 Beijing Key Laboratory for Precision Optoelectronic Measurement Instrument and Technology, Beijing 100081, China
3 Yangtze Delta Region Academy of Beijing Institute of Technology, Jiaxing 314000, China
* Correspondence: xintang@bit.edu.cn
† These authors contributed equally to this work.

Abstract: Infrared detectors with polarization sensitivity could extend the information dimension of the detected signals and improve target recognition ability. However, traditional infrared polarization detectors with epitaxial semiconductors usually suffer from low extinction ratio, complexity in structure and high cost. Here, we report a simulation study of colloidal quantum dot (CQD) infrared detectors with monolithically integrated metal wire-grid polarizer and optical cavity. The solution processibility of CQDs enables the direct integration of metallic wire-grid polarizers with CQD films. The polarization selectivity of HgTe CQDs with resonant cavity-enhanced wire-grid polarizers are studied in both short-wave and mid-wave infrared region. The extinction ratio in short-wave and mid-wave region can reach up to 40 and 60 dB, respectively. Besides high extinction ratio, the optical cavity enhanced wire-grid polarizer could also significantly improve light absorption at resonant wavelength by a factor of 1.5, which leads to higher quantum efficiency and better spectral selectivity. We believe that coupling CQD infrared detector with wire-grid polarizer and optical cavity can become a promising way to realize high-performance infrared optoelectronic devices.

Keywords: polarization; colloidal quantum dots; detectivity; optical cavity

Citation: Zhao, P.; Mu, G.; Chen, M.; Tang, X. Simulation of Resonant Cavity-Coupled Colloidal Quantum-Dot Detectors with Polarization Sensitivity. *Coatings* **2022**, *12*, 499. https://doi.org/10.3390/coatings12040499

Academic Editor: Alicia de Andrés

Received: 28 February 2022
Accepted: 5 April 2022
Published: 7 April 2022

1. Introduction

Polarization imaging technology plays an increasingly important role in target recognition and detection. The polarization state of light results from the interaction between light and objects at the microscopic level. When the surface topography, states and observation orientation are different, the specular polarization information of the object will change accordingly [1,2]. Therefore, when the temperature difference between the observed objects and the surrounding environment is relatively small, infrared thermal imaging fails to give high-contrast images, whereas polarization imaging can provide a new information dimension for target discrimination with high contrast between the targets and the surrounding background [3,4]. Therefore, polarization imaging technology has become an powerful imaging technology besides traditional intensity imaging and spectral imaging, which has a wide prospective application in military infrared anti-counterfeiting equipment [5], remote sensing, astronomical observation [6] and medical diagnosis [7].

Traditional polarization imaging techniques are mainly divided into time-sharing polarization imaging [8,9], amplitude polarization imaging and aperture polarization imaging. However, they rely on moving optics, registration and sophisticated calibration procedures, which greatly increase their complexity and cost.

With the increasing demands for high sensitivity, low cost and scalable polarized infrared detector equipment, there is a trend to directly couple the polarizers such as metal wire-grid polarizer with infrared detector on pixel level. High sensitivity, compact structure and wide bandwidth have been demonstrated with different processing technology [10,11]. Despite their high performance, these detectors are still limited by the high cost of epitaxial semiconductors [12]. It is advantageous to find a method that has both high polarization sensitivity and low cost. With the continuous progress, the maximum absorption wavelength of colloidal quantum dots (CQDs) has been extended from near infrared to longer wavelength, which provides a promising alternative for bulk infrared semiconductors such as HgCdTe, InSb and type II superlattice. Among all colloidal nano materials, mercury telluride (HgTe) CQDs has shown the highest spectral tunability so far; from short-wave infrared and mid-wave infrared to long-wave infrared and terahertz (THz) regions, we can control the reaction time and temperature to obtain colloidal quantum dots with different particle sizes for different infrared regions [13]. The solution workability and broad spectral tunability of colloidal quantum dots inspire a variety of inexpensive, high-performance optoelectronic devices [14]. Therefore, coupling metal nanowire-grid polarizer with colloidal quantum dot detector seems to be a suitable method to solve this challenge.

In this work, a high-performance polarization-sensitive infrared detector is designed by integrating HgTe CQDs photovoltaic detector with wire-grid polarizer and optical cavity. Finite element analysis of HgTe CQDs polarization-sensitive detectors was systematically conducted covering short-wave infrared and mid-wave infrared. The simulation results show that this detector has high polarization selectivity in both short-wave infrared and mid-wave infrared. The extinction ratio in short-wave and mid-wave region can reach up to 40 and 60 dB, respectively. Besides high extinction ratio, the optical cavity enhanced wire-grid polarizer could also significantly improve light absorption at resonant wavelength by a factor of 1.5, which leads to higher quantum efficiency and better spectral selectivity.

2. Modeling and Simulation

The structure of the proposed HgTe CQDs detector with resonant cavity-coupled wire-grid polarizer is illustrated in Figure 1a. Distributed Bragg reflectors (DBRs) are set on an Al_2O_3 substrate, on top of which subwavelength wire-grid polarizers are added. The reflectivity of DBR is determined by the number of the alternating layers and their difference in refractive index [15]. The polarization selectivity of the metal wire grid is originated from the structural asymmetry. The transverse wave (TE polarized light) in which the electric vector is parallel to the direction of the wire grids can excite the electrons to oscillate along the grid, so TE polarized light will be reflected, and the transverse magnetic wave (TM polarized light) with electric vector perpendicular to the wires cannot excite free electron oscillations and therefore shows high optical transmission [12]. On top of the metal wire-grid polarizer, HgTe CQDs photodiodes are constructed with indium tin oxide and gold as bottom and top contacts. The DBR and top gold contact forms a resonant cavity.

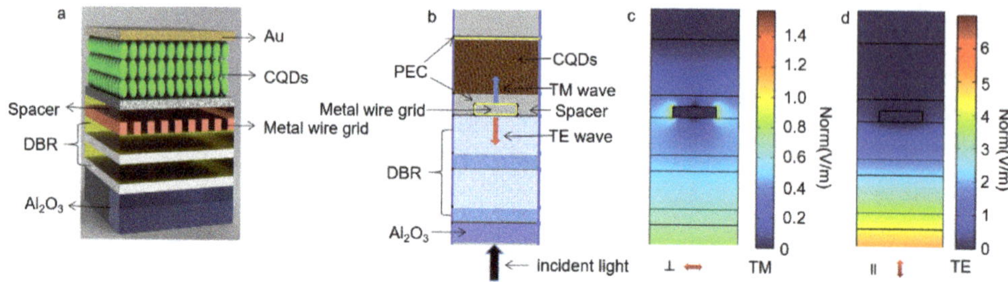

Figure 1. Architecture of resonant cavity-coupled Colloidal quantum-dot detectors with polar ization sensitivity. (**a**) Structure schematic and (**b**) simulation model of Resonant cavity-coupled Colloidal quantum-dot detectors with polarization sensitivity. Electric field distribution of (**c**) TM waves and (**d**) TE waves.

COMSOL Multiphysics is used to conduct the simulation in our study. To save computation resources, a two-dimensional (2D) simulation is used, as shown in Figure 1b. The light is incident from the bottom with controlled power of 1 W. The left and right sides are set as "Periodic boundary condition". The substrate material is chosen to be Al_2O_3 with a refractive index of 1.3 + 0i. Low refraction SiO_2 material with a refractive index of 1.4 + 0i and high refractive index Si material with a refractive index of 3.42 + 0i are selected for Bragg mirror, and SiO_2 material is also selected as optical spacer. For HgTe CQDs, the refractive index is 2.3 + 0.1i [16], and the rest of the model is set to be air with a refractive index of 1 + 0i. The wire-grid polarizer and the top electrode are set as perfect electrical conductors (PEC). The electric field distribution of TM wave and TE wave obtained by simulation is shown in Figure 1c,d, and it can be seen that the structure has strong polarization selectivity. The index of refraction of CQDs films is measured before simulation. Based on the measured index of refraction, the simulation fraction $f_{simulation}$ can be estimated by using different effective medium approximations, including the Maxwell-Garnett (MG) model and Bruggeman (BG) model:

$$\text{MGmodel}: \frac{\varepsilon_{eff} - \varepsilon_{CQD}}{\varepsilon_{eff} + 2\varepsilon_{CQD}} = (1 - f_{simulation})\frac{\varepsilon_m - \varepsilon_{CQD}}{\varepsilon_m + 2\varepsilon_{CQD}} \tag{1}$$

$$\text{BGmodel}: f_{simulation}\frac{\varepsilon_{CQD} - \varepsilon_{eff}}{\varepsilon_{CQD} + 2\varepsilon_{eff}} + (1 - f_{simulation})\frac{\varepsilon_m - \varepsilon_{eff}}{\varepsilon_m + 2\varepsilon_{eff}} = 0 \tag{2}$$

where ε_{eff} is the effective dielectric constant, which is obtained by the measured index of refraction n, ε_{CQDs} is the dielectric constant of CQDs, and ε_m is the dielectric constant of medium. For HgTe, ε_{CQDs} is taken as 15.1. ε_m is taken to be 1, since the CQDs are surrounded by air. According to the thickness of colloidal quantum dot film, the approximate refractive index is 2.3 + 0.1i; in previous studies, this value has also been proved to be reliable [14,16].

3. Results and Discussions

3.1. Effects of Period and Duty Cycle

The center wavelength λ_{center} of Bragg mirror defined the operation spectral ranges of resonant cavity, and it can be expressed as

$$\lambda_{center} = 4n_l d_l = 4n_h d_h, \tag{3}$$

where n_1 and n_h are the refractive index of the two materials, and d_1 and d_h are the thickness of the two materials. Bragg mirrors are formed by two alternating layers with different refractive index. By changing the thickness of each layer in the simulation, the reflection window can be controlled. When the response of the detector is in short-wave infrared, the center wavelength is set to be 2 μm, so the thickness of SiO_2 and Si is calculated to be 340 nm and 140 nm, respectively. When the detector works in the mid-wave region, the center wavelength is about 4 μm, and the thickness of SiO_2 and Si is set to be 700 nm and 280 nm [17].

For wire-grid polarizer, the period and duty cycle are two key parameters that influence its spectral extinction ratio [18]. The duty cycle f is defined as: $f = b/p$, where p is the period of the wire grid, and b is the width of wire grid. The polarization performance of polarizer is usually characterized by extinction ratio, that is, ER = 10 log(TM/TE). In our simulation, the extinction ratio is defined by the absorption ratio of CQDs under TM and TE wave incidence.

When short-wave infrared plane waves are incident and the period of metal wire grid increases from 0.6 μm to 1.0 μm, the spectral absorption characteristic curves of HgTe CQDs is simulated, as shown in Figure 2a. The corresponding extinction ratio of absorption is shown in Figure 2b. The HgTe CQDs demonstrate clear polarization-dependent absorption. With decreasing period from 1.0 to 0.6 μm, the extinction ration increases by almost three orders of magnitude. Duty cycle of wire grid is another important structural parameter.

When the plane wave is incident vertically, the absorption characteristic curves of HgTe CQDs with wire grids with different duty cycles are simulated in the short-wave region (Figure 2c). When the duty cycle of the grid increases from 0.3 to 0.7, the absorption rate of TM wave decreases gradually and redshifts, whereas that of TE decreases significantly, and the extinction ratio increases (Figure 2d). In the mid-wave region, when the period increase from 0.6 μm to 1.0 μm, the absorption characteristic curve of HgTe CQDs with metal wire grid is simulated, as shown in Figure 3a,b. When duty cycle changes from 0.3 to 0.7, the absorption rate of TM wave in the mid-wave region redshifts. (Figure 3c). The simulation results show that the extinction ratio of the CQDs detector in the mid-wave region is much larger than that in the short-wave region (Figure 3d). Thus, the detector will be more suitable in the mid-wave region than the short-wave region according to the simulation results.

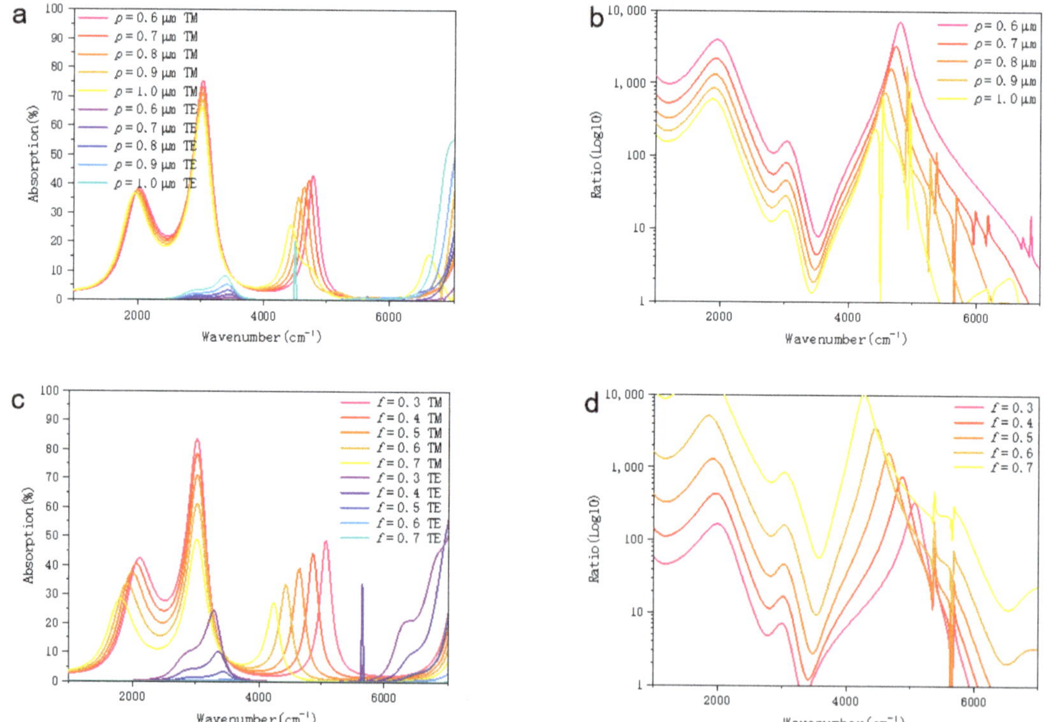

Figure 2. (**a**) Influence of different periods on polarization performance and absorption rate in short-wave region. (**b**) Extinction ratio for different periods in short-wave region. The thickness of metal wire grid is 100 nm. Duty cycle *f* is 0.5. The thickness of SiO_2 is 340 nm. The thickness Si is 140 nm. (**c**) Influence of different duty cycle on polarization performance and absorption rate in short-wave region. (**d**) Extinction ratio for different duty cycle in short-wave region. Thickness of metal wire grid is 100 nm. The period *p* is 0.8 μm. The thickness of SiO_2 is 340 nm. The thickness Si is 140 nm.

The extinction ratio in short-wave and mid-wave region according to our simulation results can reach 40 and 60 dB, respectively, through coupling the single-layer wire grid and Bragg mirror, which are higher than that of other similar structures [10,19,20]. Although the extinction ratio using multi-layers wire grid is larger than ours, the complicated structure is difficult for mass production [12]. Compared with previous studies, our structure has considerable extinction ratio on the basis of its simpler structure, which makes the detector easier to manufacture. Therefore, the coupling of colloidal quantum dot detector and nano metal wire grid will be a promising method for the development of high-performance polarization photodetectors.

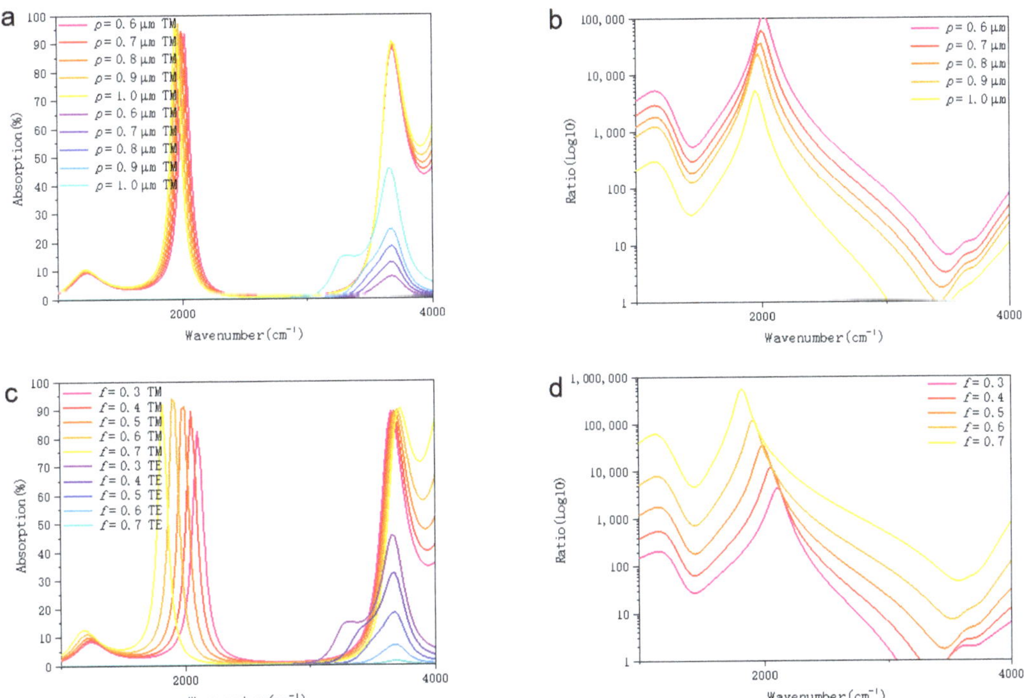

Figure 3. (**a**) Influence of different periods on polarization performance and absorption rate in mid-wave region. (**b**) Extinction ratio for different periods in mid-wave region. The thickness of metal wire grid is 100 nm. Duty cycle f is 0.5. The thickness of SiO_2 is 700 nm. The thickness Si is 280 nm. (**c**) Influence of different duty cycle on polarization performance and absorption rate in mid-wave region. (**d**) Extinction ratio for different duty cycle in mid-wave region. The thickness of the metal wire grid is 100 nm. The period p is 0.8 μm. The thickness of SiO_2 is 700 nm. The thickness of Si is 280 nm.

3.2. Effects of Optical Spacer Thickness

The optical spacer above the wire-grid polarizer is composed of SiO_2. The role of the resonator is to select the light with a desired wavelength and suppress the light with other wavelength [21]. Therefore, the desired wavelength of TM absorption peak can be selected by adjusting the thickness of the SiO_2 layer. The wavelength of the peak absorption for a resonant cavity can be expressed by

$$\lambda = \frac{2\pi l\sqrt{n^2 - \sin^2\theta}}{\pi m - \phi} \tag{4}$$

where l is the optical path, n is the index of refraction of the optical spacer, θ is the angle of incidence, m is the order of interference, and ϕ is the phase shift of light between the two reflective surfaces. Therefore, by changing the thickness of the optical spacer (SiO_2), the spectra can be tuned accordingly.

In the short-wave region and mid-wave infrared, the absorption curves of TM wave and TE wave with various thickness of the optical spacer are shown in Figure 4a,b. The black curve is the reflection window of the distributed Bragg mirror, and the TM absorption peak redshifts with the increased thickness of the optical spacer. For short-wave infrared HgTe detectors, the TM absorption peak decreases gradually with the decrease in wavelength, whereas the magnitude of absorption remains almost unchanged in the mid-wave infrared region. After the addition of optical spacer, the CQD detectors remain sensitive to the polarization direction with high absorption under TM incidence in short-wave region and mid-wave region (Figure 4c,d).

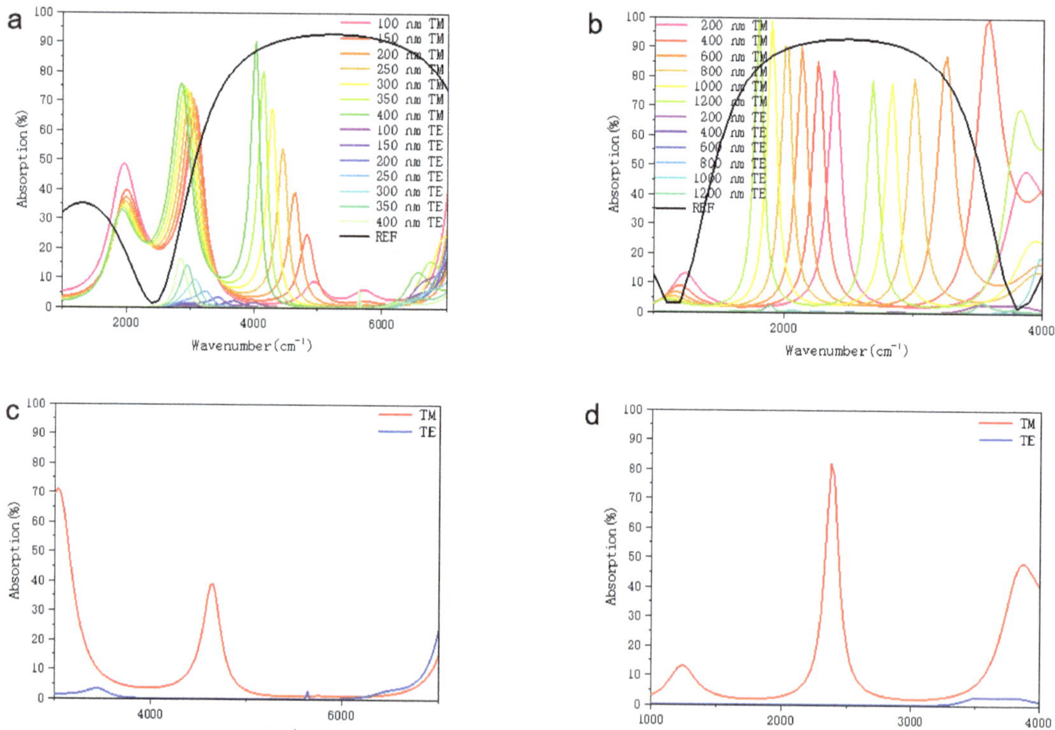

Figure 4. (**a**) TM wave and TE wave absorption curves with different resonant cavity thickness and the reflection window of Bragg mirror in short-wave region. (**b**) TM wave and TE wave absorption curves with different resonant cavity thickness and the reflection window of Bragg mirror in mid-wave region. (**c**) Single peak of TM and TE in short-wave region. (**d**) Single peak of TM and TE in mid-wave region.

3.3. Effects of Plasmon Resonance by Wire Grid

Compared to HgTe detectors without resonant cavity, the cavity can significantly improve light absorption at targeted wavelength, resulting in the enhancement of quantum efficiency [22]. Moreover, our simulation shows that the metal wire grid provides surface plasmon resonance effect which leads to further increase in the infrared absorption of detector. The absorption of light at different polarization angles by the CQDs infrared detector with resonant cavity and wire-grid polarizer in the short-wave region and mid-wave infrared is shown in Figure 5a,b. The integration of HgTe CQD photovoltaic detector, wire-grid polarizer and resonant cavity can not only provide polarization-sensitivity, but also significantly increases CQDs film absorption.

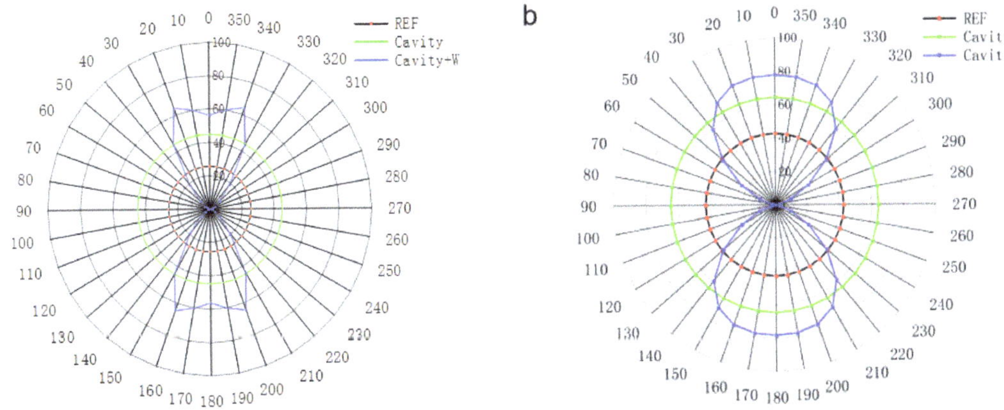

Figure 5. (**a**) Absorption effects of resonator and metal grid on different polarization angles in short-wave region. (**b**) Absorption effects of resonator and metal grid on different polarization angles in mid-wave region.

It can be seen that the proposed resonant cavity-enhanced wire-grid polarizers not only show excellent polarization selectivity, but also give enhanced absorption with high spectral selectivity, which provides a promising method to improve the spectral resolution of infrared polarization detector.

4. Conclusions

In conclusion, we proposed the design of HgTe CQDs photodetectors with resonant cavity-enhanced wire-grid polarizer, providing a theoretical study for designing low-cost polarization infrared detectors in the future. The simulation results show that the detector can work in both short-wave region and mid-wave region with high extinction ratio up to 40 dB and 60 dB. Our simulation shows that the metal wire grid provides surface plasmon resonance effect which leads to further increase in the infrared absorption of detector. Benefitting from CQDs' solution processibility, HgTe CQDs could be directed integrated with the wire-grid polarizers. Therefore, we believe that the simulation results have important implications for the focal plane array polarization imaging.

Author Contributions: Conceptualization, X.T., P.Z. and G.M.; methodology, M.C.; software, P.Z.; validation, X.T., P.Z. and G.M.; formal analysis, M.C.; investigation, X.T., P.Z. and G.M.; resources, X.T., P.Z. and G.M.; data curation, X.T., P.Z. and G.M.; writing—original draft preparation, P.Z. and G.M.; writing—review and editing, X.T. and M.C.; visualization, X.T., P.Z. and G.M. supervision, X.T.; project administration, X.T.; funding acquisition, X.T. All authors have read and agreed to the published version of the manuscript.

Funding: This research was funded by National Natural Science Foundation of China and National Key R&D Program of China (NSFC No. 62035004, 2021YFA0717600 and NSFC No. 62105022).

Institutional Review Board Statement: Not applicable.

Informed Consent Statement: Not applicable.

Data Availability Statement: Not applicable.

Conflicts of Interest: The authors declare no conflict of interest.

References

1. Tyo, J.S.; Goldstein, D.L.; Chenault, D.B.; Shaw, J.A. Review of Passive Imaging Polarimetry for Remote Sensing Applications. *Appl Opt.* **2006**, *45*, 5453–5469. [CrossRef] [PubMed]
2. Van Harten, G.; Diner, D.J.; Daugherty, B.J.S.; Rheingans, B.E.; Bull, M.A.; Seidel, F.C.; Chipman, R.A.; Cairns, B.; Wasilewski, A.P.; Knobelspiesse, K.D. Calibration and validation of Airborne Multiangle SpectroPolarimetric Imager (AirMSPI) polarization measurements. *Appl. Opt.* **2018**, *57*, 4499. [CrossRef] [PubMed]
3. Playle, N.; Port, D.; Rutherford, R.; Burch, I.; Almond, R. Infrared Polarisation Sensor for Forward Looking Mine Detection. In *Detection and Remediation Technologies for Mines and Minelike Targets VII*; SPIE: Washington, DC, USA, 2002; Volume 4742. [CrossRef]
4. Rogne, T.J.; Smith, F.G.; Rice, J.E. Passive Target Detection using Polarized Components of Infrared Signatures. *Proc. SPIE Int. Soc. Opt. Eng.* **1990**, *1317*. [CrossRef]
5. Cavanaugh, D.B.; Castle, K.R.; Davenport, W. Anomaly detection using the hyperspectral polarimetric imaging testbed. In *Proceedings of the Algorithms and Technologies for Multispectral, Hyperspectral, and Ultraspectral Imagery XII*; SPIE: Washington, DC, USA, 2006; Volume 6233, p. 62331Q. [CrossRef]
6. Crosby, F.J. Stokes vector component versus elementary factor performance in a target detection algorithm. In *Proceedings of the Polarization: Measurement, Analysis, and Remote Sensing VI*; SPIE: Washington, DC, USA, 2004; Volume 5432, p. 1. [CrossRef]
7. Liu, G.L.; Li, Y.; Cameron, B.D. Polarization-Based Optical Imaging and Processing Techniques with Application to the Cancer Diagnostics. *Proc. SPIE Int. Soc. Opt. Eng.* **2002**, *4617*. [CrossRef]
8. Gupta, N.; Denes, L.; Gottlieb, M.; Suhre, D.; Kaminsky, B.; Metes, P. Spectropolarimetric Imaging Using a Field-Portable Image. *Proc. SPIE Int. Soc. Opt. Eng.* **2000**, *4132*. [CrossRef]
9. Zhao, Y.; Gong, P.; Pan, Q. Object Detection by Spectropolarimeteric Imagery Fusion. *IEEE Trans. Geosci. Remote Sens.* **2008**, *46*, 3337–3345. [CrossRef]
10. Ekinci, Y.; Solak, H.H.; David, C.; Sigg, H. Bilayer Al wire-grids as broadband and high-performance polarizers. *Opt. Express* **2006**, *14*, 2323. [CrossRef] [PubMed]
11. Lee, K.J.; Curzan, J.; Shokooh-Saremi, M.; Magnusson, R. Resonant wideband polarizer with single silicon layer. *Appl. Phys. Lett.* **2011**, *98*, 99–102. [CrossRef]
12. Yang, Z.Y.; Lu, Y.F. Broadband nanowire-grid polarizers in ultraviolet-visible-near-infrared regions. *Opt. Express* **2007**, *15*, 9510. [CrossRef] [PubMed]
13. Keuleyan, S.; Lhuillier, E.; Guyot-Sionnest, P. Synthesis of colloidal HgTe quantum dots for narrow mid-IR emission and detection. *J. Am. Chem. Soc.* **2011**, *133*, 16422–16424. [CrossRef] [PubMed]
14. Tang, X.; Ackerman, M.M.; Guyot-Sionnest, P. Thermal Imaging with Plasmon Resonance Enhanced HgTe Colloidal Quantum Dot Photovoltaic Devices. *ACS Nano* **2018**, *12*, 7362–7370. [CrossRef] [PubMed]
15. Zhao, X.; Zhang, Y.; Zhang, Q.; Zou, B.; Schwingenschlogl, U. Transmission comb of a distributed Bragg reflector with two surface dielectric gratings. *Sci. Rep.* **2016**, *6*, 21125. [CrossRef] [PubMed]
16. Ning, Y.; Zhang, S.; Hu, Y.; Hao, Q.; Tang, X. Simulation of monolithically integrated meta-lens with colloidal quantum dot infrared detectors for enhanced absorption. *Coatings* **2020**, *10*, 1218. [CrossRef]
17. Tang, X.; Ackerman, M.M.; Guyot-Sionnest, P. Acquisition of Hyperspectral Data with Colloidal Quantum Dots. *Laser Photonics Rev.* **2019**, *13*, 1900165. [CrossRef]
18. Wu, Z.; Powers, P.E.; Sarangan, A.M.; Zhan, Q. Optical characterization of wiregrid micropolarizers designed for infrared imaging polarimetry. *Opt. Lett.* **2008**, *33*, 1653. [CrossRef] [PubMed]
19. Shirakawa, A.; Shirakawa, A.; Takaichi, K.; Takaichi, K.; Yagi, H.; Yagi, H.; Lu, J.; Lu, J.; Musha, M.; Musha, M.; et al. Diode-pumped mode-locked Yb^{3+};Y_2O_3 ceramic laser. *Opt. Express* **2003**, *11*, 2911–2916. [CrossRef] [PubMed]
20. Meng, F.; Luo, G.; Maximov, I.; Montelius, L.; Chu, J.; Xu, H. Fabrication and characterization of bilayer metal wire-grid polarizer using nanoimprint lithography on flexible plastic substrate. *Microelectron. Eng.* **2011**, *88*, 3108–3112. [CrossRef]
21. Kishino, K.; Unlu, M.S.; Chyi, J.-I.; Reed, J.; Arsenault, L.; Morkoc, H. Resonant cavity enhanced (RCE) photodetectors. *IEEE J. Quantum Electron.* **1991**, *27*, 2025–2034. [CrossRef]
22. Motogaito, A.; Morishita, Y.; Miyake, H.; Hiramatsu, K. Extraordinary Optical Transmission Exhibited by Surface Plasmon Polaritons in a Double-Layer Wire Grid Polarizer. *Plasmonics* **2015**, *10*, 1657–1662. [CrossRef]

 coatings

Article

Mid-IR Intraband Photodetectors with Colloidal Quantum Dots

Xue Zhao [1], Ge Mu [1], Xin Tang [1,2,3] and Menglu Chen [1,2,3,*]

1 School of Optics and Photonics, Beijing Institute of Technology, Beijing 100081, China; 3220210544@bit.edu.cn (X.Z.); 7520210145@bit.edu.cn (G.M.); xintang@bit.edu.cn (X.T.)
2 Beijing Key Laboratory for Precision Optoelectronic Measurement Instrument and Technology, Beijing 100081, China
3 Yangtze Delta Region Academy of Beijing Institute of Technology, Jiaxing 314019, China
* Correspondence: menglu@bit.edu.cn

Abstract: In this paper, we investigate an intraband mid-infrared photodetector based on HgSe colloidal quantum dots (CQDs). We study the size, absorption spectra, and carrier mobility of HgSe CQDs films. By regulating the time and temperature of the reaction during synthesis, we have achieved the regulation of CQDs size, and the number of electrons doped in conduction band. It is experimentally verified by the field effect transistor measurement that dark current is effectively reduced by a factor of 10 when the 1Se state is doped with two electrons compared with other doping densities. The HgSe CQDs film mobility is also measured as a function of temperature the HgSe CQDs thin film detector, which could be well fitted by Marcus Theory with a maximum of 0.046 ± 0.002 cm^2/Vs at room temperature. Finally, we experimentally discuss the device performance such as photocurrent and responsivity. The responsivity reaches a maximum of 0.135 ± 0.012 A/W at liquid nitrogen temperature with a narrow band photocurrent spectrum.

Keywords: intraband photodetector; colloidal quantum dot; HgSe; mid-infrared

Citation: Zhao, X.; Mu, G.; Tang, X.; Chen, M. Mid-IR Intraband Photodetectors with Colloidal Quantum Dots. *Coatings* **2022**, *12*, 467. https://doi.org/10.3390/coatings12040467

Academic Editor: Ana-Maria Lepadatu

Received: 28 February 2022
Accepted: 23 March 2022
Published: 30 March 2022

1. Introduction

In recent years, colloidal quantum dots (CQDs) devices have gradually been widely used in imaging, remote sensing, gas detection, medical diagnosis, and epidemic prevention and control [1–4]. The performance of CQDs photodetectors has been comparable to commercial InGaAs photodiodes but order-of-magnitude reduction in cost [5]. Compared traditional infrared photodetectors, CQDs photodetectors theoretically offer the advantages of higher operating ambient temperatures, smaller dark currents, faster response performance [6], and lower fabrication cost [7].

The materials selected for the infrared photodetectors with interband transition [8,9] are usually limited in semimetal or ternary alloy with a small gap [10]. However, the intraband CQD devices, utilizing optical transitions from lowest electron state labelled 1Se and the next level 1Pe, hold unique promise to extend the wavelength from near- and mid-infrared to long-wave infrared (Supplementary Materials Figure S1). What is more, intraband photodetectors promise a better photoelectric performance [11,12] due to the low conduction band density which inhibit the process of Auger relaxation [13–16]. For example, the Auger relaxation coefficient is at least three orders smaller in intraband HgSe CQDs compared with the bulk materials with the same energy gap [17]. Hence, the study of CQDs intraband photodetectors has important research significance. The air stable n-type mercury (ii) selenide (HgSe) CQDs provide an excellent platform for studying intraband photodetectors [18,19].

In current stage, due to the limited study on intraband materials, the intraband photodetectors based in HgSe CQDs have excessive dark current compared to interband CQD photodetectors at the same wavelength. The solution is to precisely adjust the number of electrons in 1Se [11,20]. The unpaired electrons in Se state or Pe state would result in large dark current. The dark current is minimal when 1Se is fully filled which is two electrons in conduction band per CQD (Supplementary Materials).

To address these challenges, in this paper, we provide a way to control the doping inside the HgSe CQDs, which is characterized by optical and transport measurements. Then, we build the HgSe CQDs intraband photodetector in mid-infrared (Mid-IR) and further analyze its photocurrent, and responsivity.

2. Materials and Methods

In a glove box with a nitrogen environment, 27.2 mg of mercury chloride ($HgCl_2$) and 4 mL of oleylamine (OAm) were placed in a 40 mL vial, and were heated at 100 °C for 1 h, and then were heat-balanced at 115 °C (or 120 °C) for half an hour. A total of 126 mg (0.1 mmol) of selenium urea was dissolved in 10 mL of OAm and heated at 180 °C under nitrogen for 2 h to form a brown transparent liquid. A total of 1 mL of this solution was taken and injected into the mercury chloride/oleylamine ($HgCl_2$/OAm) solution, and was reacted at 115 °C for 3–8 min. We removed the reaction solution out of the glove box and used cold water to end the reaction. After the end of the reaction, the HgSe CQDs solution was placed in a centrifuge tube. A total of 1 mL of 1-dodecyl mercaptan (DDT), 20 drops of didecyldimethylammonium bromide (DDAB), and 15 mL of methanol (MeOH) were added to the centrifuge tube. The solution in the centrifuge tube was centrifuged at 7500 r.p.m for 6 min. After centrifugation, the supernatant was decanted, and the remaining pellet was dissolved in chlorobenzene and filtered through a disposable filter.

The photocurrent spectrum of the material with Nicolet iS20 FTIR (Thermo Fisher Scientific, Waltham, MA, USA) was measured. The FTIR implemented 500,000:1 signal-to-noise ratio and 0.25 cm^{-1} spectral resolution.

Photoconductance spectra was measured with a standard Thermo Scientific Nicolet iS20 FTIR Spectrometer. The internal light was reflected in the sample with a gold mirror. The samples were biased with a 3 V battery in series. The current across the sample was sent to transimpedance amplifier Femto-200 with a gain of 10,000 before being sent back to the FTIR input.

3. Results and Discussion

3.1. Absorption Spectra

The Transmission Electron Microscope (TEM) image of HgSe CQDs with a synthesis temperature of 115 °C and a synthesis time of 5 min is shown in Figure 1a. The average size of CQDs is 4.7 nm and the size distribution is uniform with the standard deviation of 0.5 nm (Supplementary Materials Figure S2). As can be seen in Figure 1b, the size of the CQDs can be varied by changing the synthesis temperature and time [21,22]. Changing the size of CQDs can achieve the movement of the absorption spectrum. The intraband absorption peak of HgSe CQDs is between 2000–3000 cm^{-1}, and the absorption peak will be redshifted by increasing the reaction time. The relationship between CQDs of different sizes and energy bands is shown in Figure 1c. When the size of CQDs is "small", the Fermi level is lower than Se states, and the HgSe CQDs are defined as "n^- type". When the size of CQDs is "medium", the energy gaps become smaller, and the band shift. As a result, the Fermi level is in Se state, which we define as two electrons in conduction band per CQD [17]. When the size of CQDs is "large", the Fermi level is higher than Se states, and the HgSe CQDs are defined as "n^+ type". By changing the size of CQDs, the position of the absorption peaks in intraband and interband can be changed in Figure 1b. HgSe intraband photodetector can specifically detect signals in the Mid-IR range, possessing potential application in gas detection such as CO_2 with a signature absorption of 2400 cm^{-1}.

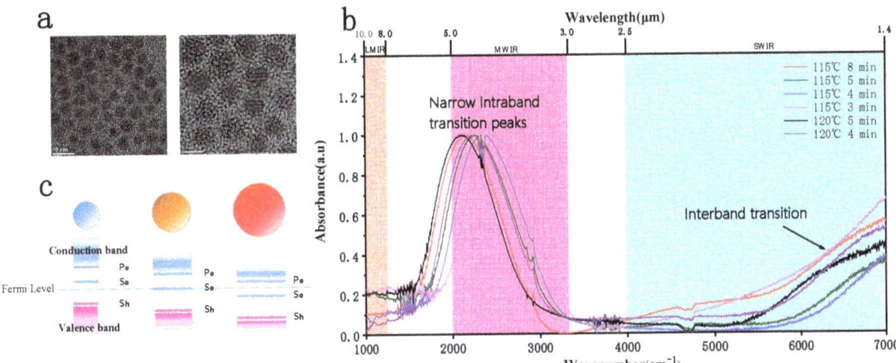

Figure 1. Characterization of HgSe CQDs materials. (**a**) TEM images of HgSe CQDs with a particle diameter of 4.7 nm and standard deviation of 0.5 nm. (**b**) Absorption spectra of HgSe CQDs. (**c**) Diagram of the relationship between the size of CQDs and energy bands.

3.2. Transport Properties of HgSe CQDs

Transport properties such as carrier type, carrier density, and carrier mobility greatly affect the photodetector performance. We can judge the performance of the material by measuring the carrier type and mobility of the HgSe film [23]. This information is given by the field effect transistor (FET) measurement. HgSe CQDs film is spin-coated on 300 nm SiO_2/Si substrate, which has a layer of interdigited electrodes. A solid ligands exchange process with an ethanedithiol/hydrochloric acid (EDT/HCl) solution is followed. In this process, short ligands would displace the long chain OAm ligands between CQDs. After the ligand exchange, we rinse the film with isopropanol (IPA).

The HgSe/EDT CQD film is measured by a field effect transistor circuit. Whether CQDs are "n^- type" or "n^+ type" can be determined by the slope sign of the FET transfer curves around zero gate potential. A schematic diagram of the FET structure is shown in Figure 2a.

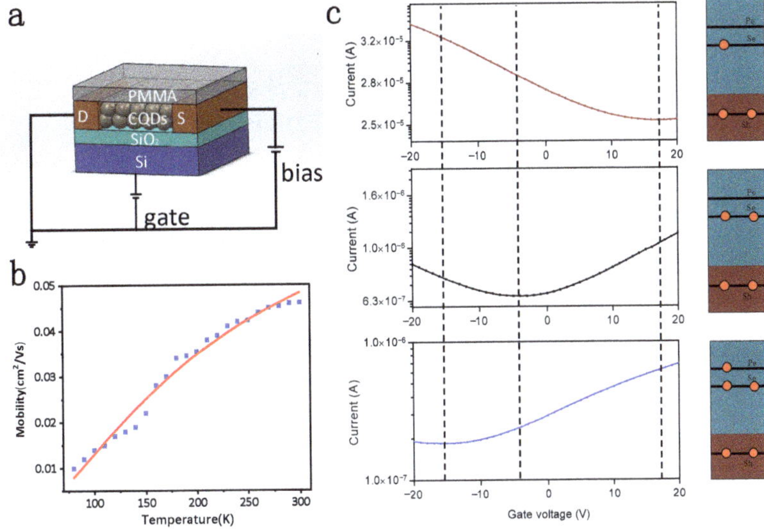

Figure 2. Transport study on HgSe CQDs. (**a**) Schematic diagram of the field effect pipe structure. (**b**) Carrier mobility at different temperatures and mobility fitted with Marcus theory as a function of temperature. (**c**) FET source-drain current as a function of voltage for HgSe/EDT at 80 K.

For CQDs photodetectors, dark current and noise are important figures of merit. The shot noise has a great influence on the CQDs photoconductivity detectors, and the formula for shot noise is shown in Equation (1):

$$i_n = \sqrt{4eI_{dark}} \tag{1}$$

where e is the electron charge, I_{dark} is the dark current of the photodetector.

The dark current is an important factor affecting the noise in Equation (1) and the root cause is the thermally excited electron-hole pairs. Therefore, the dark current has an exponential relationship with the temperature [24]. For photodetectors with intraband transition, the dark current was minimal at low temperature [25]. Therefore, after using liquid nitrogen to cool down to 80 K, the dark current is effectively reduced. The curve of FET source-drain current as a function of voltage for HgSe/EDT at 80 K is shown in Figure 2c.

In addition to cooling down to reduce the dark current, the effect of the dark current can also be reduced by adjusting the number of electrons filled in the 1Se band. By varying the temperature and time of synthesis, the number of electrons filled by HgSe CQDs in the 1Se band can be regulated. In Figure 2c, when the number of doped electrons is two, the dark current is minimal. Compared with "n^- type" or "n^+ type" CQDs, the CQDs with fully doped Se state have a ten-fold reduction in dark current.

The mobility of HgSe/EDT thin films can be calculated through the FET circuit, and the mobility calculation formula is as follows [26]:

$$\mu_{FET} = \frac{L}{WC_i V_D} \bullet \frac{\Delta I_D}{\Delta V_G} \tag{2}$$

Among them, the thickness of silica (SiO_2) is 300 nm, $C_i = 1.15 \times 10^{-4}\ F/m^2$ is the capacitance, $L = 10\ \mu m$ is the gap, $W = 1000\ \mu m \times 50$ is total channel width, "50" is the number of channels, and $V_D = 1\ V$ is drain voltage. $\frac{\Delta I_D}{\Delta V_G}$ is the slope of the transfer current and the gate voltage. After calculation, the mobility of HgSe films exchanged with EDT can reach $0.046 \pm 0.002\ cm^2/Vs$ at room temperature.

Figure 2b shows the mobility curve with temperature. The low temperature limits the movement of the carrier, and the carrier mobility decreases to $0.012 \pm 0.001\ cm^2/Vs$ in Figure 2b. The increase in temperature will lead to less binding of the carrier and an upward trend in the mobility rate. After 260 K, the migration rate's growth rate slows down. The highest mobility is $0.046 \pm 0.002\ cm^2/Vs$ at room temperature. The trend of migration change can be explained by Marcus theory:

$$\mu_{FET} = Con\,\exp(-(\gamma + \Delta G)^2/4\gamma k_B T) \tag{3}$$

where Con is the constant, ΔG the energy difference caused by uneven size of CQDs, $k_B = 1.380649 \times 10^{-23} \frac{J}{K}$ is the Boltzmann constant. The polarization effect of the medium around the carrier is given by the recombinant energy γ [27]. γ is expressed by the following equation:

$$\gamma = \frac{e}{4\pi\varepsilon_0}\left(\frac{1}{a} - \frac{1}{2(a+b)}\right)\left(\frac{1}{\varepsilon_1} - \frac{1}{\varepsilon_{st}}\right) \tag{4}$$

where $e = 1.6 \times 10^{-19}$C is the elementary charge, $\varepsilon_0 = 8.85 \times 10^{-12} \frac{F}{m}$ is the vacuum permittivity, $a = 2.3\ nm$ is the radius of the CQDs, b = 1 nm is the spacing between CQDs, $\varepsilon_1 = 6.8$ is the optical dielectric constant of matrix surrounding CQDs which is measured by the ellipsometry, $\varepsilon_{st} = 10.9$ is the static dielectric constant. We can get $\gamma = 16\ meV$, G = 16.9 meV and $Con = 0.09$ by the Marcus fitting calculation. The disorder energy is similar to the values in the previous reference [26].

3.3. Photocurrent

The photocurrent and dark current curves of the HgSe CQDs intraband photodetector are shown in Figure 3a. As shown in Figure 3a, the $\frac{I_{photocurrent}}{I_{darkcurrent}} = 10\%$ when the bias voltage is ±3 V. Compared with previous studies [28], our work effectively improves the light–dark current ratio of pure HgSe quantum dots thin films. Figure 3b shows the photocurrent spectrum of the HgSe detector at 80 K. The photocurrent spectrum of the HgSe intraband photodetector is narrow at 80 K and the photoconduction is from 2000–3000 cm^{-1} in Figure 3b. The HgSe photodetector has a narrow photoresponse, making it possible to react more sensitively to change in the outside world. And the HgSe intraband transition photodetector has an obvious photocurrent at low temperatures, so it can be widely used in harsh low temperature environments for detection.

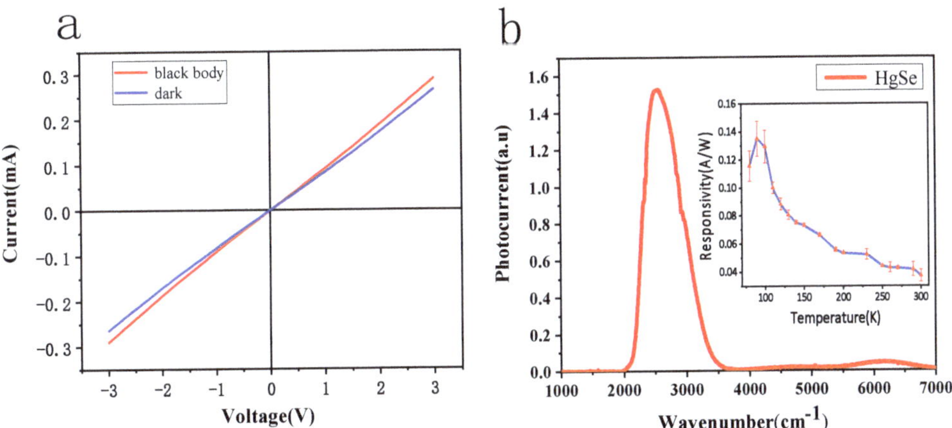

Figure 3. Photocurrent on HgSe CQDs. (**a**) HgSe CQDs as a function of current and voltage at 80 K. The red line represents the photocurrent when photodetector is illuminated by blackbody, and the blue line represents the dark current. (**b**) Photocurrent spectra with HgSe CQDs films at 80 K. The right inset graph shows the responsivity of HgSe CQDs thin films as a function of temperature.

The responsivity is calculated by measuring the photocurrent of the HgSe photoconductor device. The right-inset graph in Figure 3b shows the responsivity of HgSe thin films as a function of temperature (Supplementary Materials). It can be seen from the figure that the responsivity reaches a maximum of 0.135 ± 0.012 A/W at 90 K. It gradually heats up after 90 K, and the responsivity gradually decreases.

HgSe CQDs intraband photodetectors have narrow band spectral response in 2000–3000 cm^{-1}, which can be used in gas detection. Due to the detection capability of the intraband photodetectors in the infrared field, it can be applied to the fields of infrared imaging, target recognition, and radar detection, and other fields. Through the study of HgSe CQDs intraband photodetectors, we found that HgSe CQDs intraband photodetectors have better photoelectric performance at low temperatures. Therefore, HgSe CQDs intraband photodetectors can be applied to infrared detection in a low temperature environment. HgSe CQDs intraband photodetectors have broad application prospects.

4. Conclusions

In summary, this paper mainly introduces the intraband photodetector based on HgSe CQDs. Its intraband absorption spectrum is between 2000–3000 cm^{-1}. We provide a method to achieve accurate electron doping through repeated experiments to suppress the dark currents. When the fully doped Se state is obtained experimentally, the effect of the dark current is minimal. Compared with "n^{-} type" and "n^{+} type" CQDs, our dark current is effectively reduced by a factor of 10 measured by the field effect transistor. The mobility of HgSe/EDT films measured by FET can reach 0.046 ± 0.002 cm^2/Vs. We experimentally calculate the responsivity of HgSe CQDs to be 0.135 ± 0.012 A/W. Through the study of the performance of the intraband infrared photodetector, the selection of infrared photodetector materials can be broadened, which is of great significance for infrared detection.

Supplementary Materials: The following are available online at https://www.mdpi.com/article/10.3390/coatings12040467/s1 [29,30]. Figure S1. Bands diagram; Figure S2. TEM image and size analysis of the 5 mins sample.

Author Contributions: Conceptualization, X.Z., X.T. and M.C.; Formal analysis, X.Z. and X.T.; Funding acquisition, M.C.; Investigation, G.M.; Methodology, X.Z.; Project administration, M.C.; Supervision, X.T. and M.C.; Validation, M.C.; Visualization, X.T.; Writing—original draft, X.Z.; Writing—review & editing, X.Z., G.M., X.T. and M.C. All authors have read and agreed to the published version of the manuscript.

Funding: This work is funded by NSFC 62105022, Beijing Nova Program Z211100002121069.

Institutional Review Board Statement: Not applicable.

Informed Consent Statement: Not applicable.

Data Availability Statement: Not applicable.

Conflicts of Interest: The authors declare no conflict of interest.

References

1. Lhuillier, E.; Guyot-Sionnest, P. Recent Progresses in Mid Infrared Nanocrystal based Optoelectronics. *IEEE J. Sel. Top. Quantum Electron.* **2017**, *23*, 6000208. [CrossRef]
2. Livache, C.; Goubet, N.; Gréboval, C.; Martinez, B.; Ramade, J.; Qu, J.; Triboulin, A.; Cruguel, H.; Baptiste, B.; Klotz, S.; et al. Effect of Pressure on Interband and Intraband Transition of Mercury Chalcogenide Quantum Dots. *J. Phys. Chem. C* **2019**, *123*, 13122–13130. [CrossRef]
3. Phillips, J. Evaluation of the fundamental properties of quantum dot infrared detectors. *J. Appl. Phys.* **2002**, *91*, 4590–4594. [CrossRef]
4. Kershaw, S.V.; Susha, A.S.; Rogach, A.L. Narrow bandgap colloidal metal chalcogenide quantum dots: Synthetic methods, heterostructures, assemblies, electronic and infrared optical properties. *Chem. Soc. Rev.* **2013**, *42*, 3033–3087. [CrossRef]
5. Malinowski, P.E.; Georgitzikis, E.; Maes, J.; Vamvaka, I.; Frazzica, F.; Van Olmen, J.; De Moor, P.; Heremans, P.; Hens, Z.; Cheyns, D. Thin-Film Quantum Dot Photodiode for Monolithic Infrared Image Sensors. *Sensors* **2017**, *17*, 2867. [CrossRef]
6. Martyniuk, P.; Rogalski, A. Quantum-dot infrared photodetectors: Status and outlook. *Prog. Quantum Electron.* **2008**, *32*, 89–120. [CrossRef]
7. Keuleyan, S.; Lhuillier, E.; Guyot-Sionnest, P. Synthesis of Colloidal HgTe Quantum Dots for Narrow Mid-IR Emission and Detection. *J. Am. Chem. Soc.* **2011**, *133*, 16422–16424. [CrossRef] [PubMed]
8. Keuleyan, S.; Lhuillier, E.; Brajuskovic, V.; Guyot-Sionnest, P. Mid-infrared HgTe colloidal quantum dot photodetectors. *Nat. Photonics* **2011**, *5*, 489–493. [CrossRef]
9. Guyot-Sionnest, P.; Roberts, J.A. Background limited mid-infrared photodetection with photovoltaic HgTe colloidal quantum dots. *Appl. Phys. Lett.* **2015**, *107*, 253104. [CrossRef]
10. Deng, Z.; Jeong, K.S.; Guyot-Sionnest, P. Colloidal Quantum Dots Intraband Photodetectors. *ACS Nano* **2014**, *8*, 11707–11714. [CrossRef] [PubMed]
11. Livache, C.; Martinez, B.; Goubet, N.; Greboval, C.; Qu, J.; Chu, A.; Royer, S.; Ithurria, S.; Silly, M.G.; Dubertret, B.; et al. A colloidal quantum dot infrared photodetector and its use for intraband detection. *Nat. Commun.* **2019**, *10*, 2125. [CrossRef]
12. Ramiro, I.; Ozdemir, O.; Christodoulou, S.; Gupta, S.; Dalmases, M.; Torre, I.; Konstantatos, G. Mid- and Long-Wave Infrared Optoelectronics via Intraband Transitions in PbS Colloidal Quantum Dots. *Nano Lett.* **2020**, *20*, 1003–1008. [CrossRef]
13. Li, S.; Xia, J. Intraband optical absorption in semiconductor coupled quantum dots. *Phys. Rev. B Condens. Matter* **1997**, *55*, 15434–15437. [CrossRef]
14. Sato, S.A.; Lucchini, M.; Volkov, M.; Schlaepfer, F.; Gallmann, L.; Keller, U.; Rubio, A. Role of intraband transitions in photocarrier generation. *Phys. Rev. B* **2018**, *98*, 035202. [CrossRef]
15. Qu, J.; Goubet, N.; Livache, C.; Martinez, B.; Amelot, D.; Gréboval, C.; Chu, A.; Ramade, J.; Cruguel, H.; Ithurria, S.; et al. Intraband Mid-Infrared Transitions in Ag_2Se Nanocrystals: Potential and Limitations for Hg-Free Low-Cost Photodetection. *J. Phys. Chem. C* **2018**, *122*, 18161–18167. [CrossRef]
16. Park, M.; Choi, D.; Choi, Y.; Shin, H.B.; Jeong, K.S. Mid-Infrared Intraband Transition of Metal Excess Colloidal Ag_2Se Nanocrystals. *ACS Photonics* **2018**, *5*, 1907–1911. [CrossRef]

17. Melnychuk, C.; Guyot-Sionnest, P. Auger Suppression in n-Type HgSe Colloidal Quantum Dots. *ACS Nano* **2019**, *13*, 10512–10519. [CrossRef]
18. Chen, M.; Guyot-Sionnest, P. Reversible Electrochemistry of Mercury Chalcogenide Colloidal Quantum Dot Films. *ACS Nano* **2017**, *11*, 4165–4173. [CrossRef]
19. Martinez, B.; Livache, C.; Notemgnou Mouafo, L.D.; Goubet, N.; Keuleyan, S.; Cruguel, H.; Ithurria, S.; Aubin, H.; Ouerghi, A.; Doudin, B.; et al. HgSe Self-Doped Nanocrystals as a Platform to Investigate the Effects of Vanishing Confinement. *ACS Appl. Mater. Interfaces* **2017**, *9*, 36173–36180. [CrossRef]
20. Chen, M.; Shen, G.; Guyot-Sionnest, P. Size Distribution Effects on Mobility and Intraband Gap of HgSe Quantum Dots. *J. Phys. Chem. C* **2020**, *124*, 16216–16221. [CrossRef]
21. Grigel, V.; Sagar, L.K.; De Nolf, K.; Zhao, Q.; Vantomme, A.; De Roo, J.; Infante, I.; Hens, Z. The Surface Chemistry of Colloidal HgSe Nanocrystals, toward Stoichiometric Quantum Dots by Design. *Chem. Mater.* **2018**, *30*, 7637–7647. [CrossRef]
22. Martinez, B.; Livache, C.; Meriggio, E.; Xu, X.Z.; Cruguel, H.; Lacaze, E.; Proust, A.; Ithurria, S.; Silly, M.G.; Cabailh, G.; et al. Polyoxometalate as Control Agent for the Doping in HgSe Self-Doped Nanocrystals. *J. Phys. Chem. C* **2018**, *122*, 26680–26685. [CrossRef]
23. Lan, X.; Chen, M.; Hudson, M.H.; Kamysbayev, V.; Wang, Y.; Guyot-Sionnest, P.; Talapin, D.V. Quantum dot solids showing state-resolved band-like transport. *Nat. Mater.* **2020**, *19*, 323–329. [CrossRef]
24. Mahmoodi, A.; Dehdashti Jahromi, H.; Sheikhi, M.H. Dark Current Modeling and Noise Analysis in Quantum Dot Infrared Photodetectors. *IEEE Sens. J.* **2015**, *15*, 5504–5509. [CrossRef]
25. Liu, H.; Zhang, J.; Ackerman, M.M. Dark current and noise analyses of quantum dot infrared photodetectors. *Appl. Opt.* **2012**, *51*, 2767–2771. [CrossRef] [PubMed]
26. Chen, M.; Shen, G.; Guyot-Sionnest, P. State-Resolved Mobility of 1 $cm^2/(Vs)$ with HgSe Quantum Dot Films. *J. Phys. Chem. Lett.* **2020**, *11*, 2303–2307. [CrossRef]
27. Prodanovic, N.; Vukmirovic, N.; Ikonic, Z.; Harrison, P.; Indjin, D. Importance of Polaronic Effects for Charge Transport in CdSe Quantum Dot Solids. *J. Phys. Chem. Lett.* **2014**, *5*, 1335–1340. [CrossRef]
28. Izquierdo, E.; Dufour, M.; Chu, A.; Livache, C.; Martinez, B.; Amelot, D.; Patriarche, G.; Lequeux, N.; Lhuillier, E.; Ithurria, S. Coupled HgSe Colloidal Quantum Wells through a Tunable Barrier: A Strategy to Uncouple Optical and Transport Band Gap. *Chem. Mater.* **2018**, *30*, 4065–4072. [CrossRef]
29. Chen, M.; Lan, X.; Tang, X.; Wang, Y.; Hudson, M.H.; Talapin, D.V.; Guyot-Sionnest, P. High Carrier Mobility in HgTe Quantum Dot Solids Improves Mid-IR Photodetectors. *ACS Photonics* **2019**, *6*, 2358–2365. [CrossRef]
30. Tang, X.; Ackerman, M.M.; Guyot-Sionnest, P. Thermal imaging with plasmon resonance enhanced HgTe colloidal quantum dot photovoltaic devices. *ACS Nano* **2018**, *12*, 7362–7370. [CrossRef] [PubMed]

MDPI
St. Alban-Anlage 66
4052 Basel
Switzerland
www.mdpi.com

Coatings Editorial Office
E-mail: coatings@mdpi.com
www.mdpi.com/journal/coatings

www.ingramcontent.com/pod-product-compliance
Lightning Source LLC
LaVergne TN
LVHW070844170726
843515LV00005B/571

* 9 7 8 3 0 3 6 5 9 0 1 8 9 *